普通高等教育一流本科专业建设成果教材

仿生科学与工程专业英语

English for Bionic Science and Engineering

郭　丽　钱志辉　主编

化学工业出版社
·北京·

内容简介

《仿生科学与工程专业英语》教材以仿生学原理及其多学科应用方向为基础，参考国外原版课程素材和国际高水平学术期刊论文，系统介绍仿生相关学会组织与机构、仿生设计原则与方法、运动仿生学、仿生材料、仿生机械、仿生健康和仿生智能等内容。本教材面向国家建设现代化教育强国的战略需求，落实立德树人根本任务，通过提高仿生科学与工程专业领域的语言知识水平及综合运用能力，强化提升学生的国际视野和综合素养，培养学生成为我国强有力参与仿生科技领域国际事务与国际竞争、建设现代化科技强国的优秀人才。

本教材不仅适用于仿生科学与工程专业本科生教学，也可作为该专业硕博研究生专业英语的参考用书，还可作为材料科学与工程、机械设计制造及其自动化、人工智能等专业英语教学和自学参考用书。

图书在版编目（CIP）数据

仿生科学与工程专业英语 / 郭丽，钱志辉主编.
—北京：化学工业出版社，2023.8
普通高等教育一流本科专业建设成果教材
ISBN 978-7-122-43792-1

Ⅰ.①仿… Ⅱ.①郭… ②钱… Ⅲ.①仿生学-英语-高等学校-教材 Ⅳ.①Q811

中国国家版本馆 CIP 数据核字（2023）第 126385 号

责任编辑：傅四周　　文字编辑：王　硕
责任校对：宋　玮　　装帧设计：张　辉

出版发行：化学工业出版社（北京市东城区青年湖南街 13 号　邮政编码 100011）
印　　装：北京天宇星印刷厂
787mm×1092mm　1/16　印张 10　字数 196 千字　2023 年 10 月北京第 1 版第 1 次印刷

购书咨询：010-64518888　　售后服务：010-64518899
网　　址：http://www.cip.com.cn
凡购买本书，如有缺损质量问题，本社销售中心负责调换。

定　　价：39.00 元

前 言

本书是为高等学校仿生科学与工程专业学生学习专业英语所编写的教材。仿生科学与工程是国家新工科重点建设一级学科，是涵盖机械工程、材料科学、生物学、控制科学、力学、医学等的一门交叉学科。吉林大学仿生科学与工程专业是国家一流本科重点建设专业，旨在培养适应社会与科技发展需要并具备创新精神、实践能力和国际视野的交叉复合型高素质人才。因此，开设仿生科学与工程专业英语课程的目的是扩大学生在交叉学科领域的专业英语词汇量，培养学生阅读和翻译仿生相关专业英语文献的能力，拓展学生学习仿生专业知识的渠道，培养学生成为我国强有力参与仿生科技领域国际事务与国际竞争、建设现代化科技强国的优秀人才。

本教材面向国家建设现代化教育强国的战略需求，落实立德树人根本任务，以仿生学原理及其应用方向为基础，系统地介绍仿生专业英语的特点、仿生索引关键词、仿生相关学会组织与机构、仿生设计原则与方法、运动仿生学、仿生材料、仿生机器人和人工智能等内容，注重基础课程和专业课程的融合，提升读者在仿生科学与工程专业领域的语言知识水平及综合运用能力，为学生后续在工作学习中获取仿生专业英语信息、掌握学科发展动态、增强与国际同行交流沟通能力等奠定坚实的基础。教材参考国外原版课程素材和国际高水平学术期刊论文，选取内容知识结构完整、难度适中，并对常用和生僻的专业词汇进行了注释（专业性较强的词汇未标注音标）供读者参考。教材不仅适用于仿生科学与工程专业本科教学，也可用于硕博研究生专业英语的参考用书，或作为材料科学与工程、机械设计制造及其自动化、人工智能等专业英语教学和自学参考用书。

本教材由吉林大学郭丽和钱志辉教授主编，其中郭丽主要负责第1~4章的编写，钱志辉负责第5～7章的编写，邹猛教授进行审稿校正。编写过程中，除了引用英文原版书籍和学术论文素材，也参考了仿生学领域相关网站上的资料内容，并借鉴了其他学科专业英语编写纲要和形式，在此特向所有相关文献的作者表示诚挚的感谢。

由于编者水平有限，书中难免有疏漏之处，恳请读者、同行和专家批评指正并提出宝贵意见！

编者

2023年7月

Contents

Chapter 1

Introduction

1.1 Features of English for Special Purposes

English for Special Purposes (ESP) is a subset of English as a second or foreign language. It usually refers to teaching the English language to university students or people already in employment, with reference to the particular vocabulary and skills they need. Generally, advanced students who have a specific area of academic or professional interest should consider these programs. As with any language taught for specific purposes, a given course of ESP will focus on one occupation or profession. English for academic purposes, taught to students before or during their degrees, is one sort of ESP.

English for Bionic Science and Engineering (BSE) differs from general English language courses and contains the following characteristics:

① Designed to meet the specific needs of junior college students or graduate students majored in BSE or related majors;

② Related in themes and topics to the discipline and occupation of BSE;

③ Makes use of underlying methodology and activities of the discipline it serves;

④ Uses authentic work-specific documents and materials;

⑤ Restricted in the academic and communication skills to be learned;

⑥ Promotes cultural awareness and seeks to improve intercultural competency;

⑦ Delivers intermediate and advanced level language training specific to the discipline and occupation of BSE.

1.2 Introduction to Bionics

During the past years, bionics, a new discipline, which is dedicated to the transfer of principles of construction, regulation, interaction and organization in biology into innovative technical solutions, is attracting significant interest from various industries. The already wide spread, but still increasing application of bionic concepts and strategies in product and process development will lead to an increased request for engineers, who have a solid background knowledge in this field, accompanied by a consistent practical experience in the application of the tools and ideas offered by bionics. In order to understand the general approach of bionics and its benefit for modern engineering, a short introduction into bionics will be presented first.

1.2.1 Learning from Nature

Darwin's discovery of the evolution and adaption of living structures to their environment was one of the most shocking events in the history of human science and culture. The fact that we are not an extraordinary species standing high above the rest of our cohabitants, but that we are only one of many modifications of one basic design which had been developing for more than 3 billions of years, caused widespread and heated debates. On the other hand, people more open to scientific understanding soon learned that a powerful tool was inherent within the principles of species crossing, mutation, selection, and adaption. Some of them were thus lead to observe nature's design process and to attempt to adapt and copy it to their specific tasks. They learned that some feathers at the end of a bird's wing improved the stability at higher flight velocities [Fig.1.1(a)], different types of eyes made different species fit to recognize prey or predator, and smooth connections between trees trunks and branches reduced the danger of breaking [Fig.1.1(b)]. Many other examples are known, such as in the organization of social systems from human communities to ant colonies which perform efficient travels to keep their large population fed and housed. Looking at nature from a designer's point of view, we see nothing but the preliminary results of never-ending optimization processes.

Fig.1.1 Examples of optimization in engineering and nature. (a) Winglet of an airplane inspired by birds of prey. (b) Connection of branches to a trunk

We understand today that, from a certain point of view, all living beings are the results of optimization processes. Common to all these processes is that they have been ongoing for a very long time, using many cycles and producing many less successful variants. However, the different species would not have been able to survive, if they had not been able to adapt to the ever-changing environment.

Bionics can be seen as a relatively new field eventually contributing to the attainment of sustainable development as an "instrument provider" and a gate opener to an era of new technologies. Actually, biodiversity provides the "evolved blueprints" for bionics.

Life on planet earth, and hence biodiversity, is a product of over 4 billion years of evolution. Surprisingly little, however, is known about biodiversity. Some 1.8 million organisms have been described scientifically. Currently, efforts are underway to present a comprehensive tree of life. More than 80% of all species—based on the conventional data—are unknown and major groups of organisms are only poorly understood. Based on a 10 million total estimate, arthropods make up around 5 million (50%) of all species (Fig.1.2). However, total terrestrial arthropod species have been estimated at 6.8 million, with a range of 5.9-7.8 million. This contrasts with the much better understood vertebrates, of which only around 66,500 species are known to science. Terrestrial (vascular) plants, comprising the flowering plants, gymnosperms and ferns, are comparatively poor in species numbers. These are estimated at only 370,000 species. Plants have been well studied for centuries and, as a result, the percentage of total species numbers known (over 90%) is extremely high compared with other taxonomic groups. More recently, it was estimated that flowering plants comprise 450,000 species.

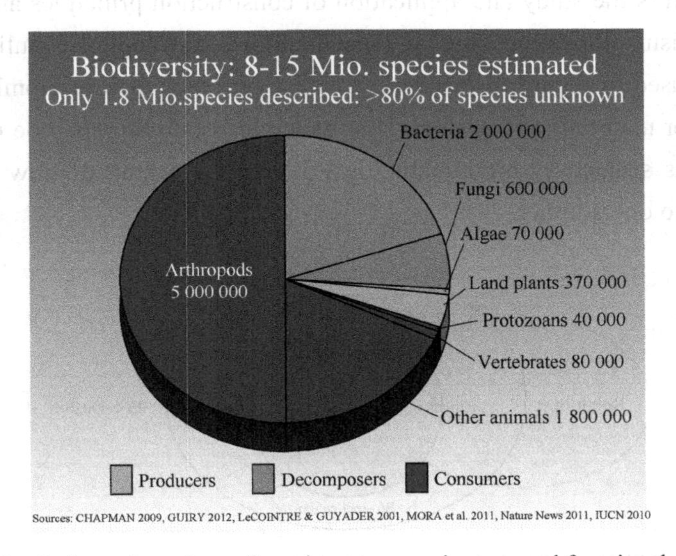

Fig.1.2 Estimated numbers of species among major taxa and functional groups

Biodiversity is the inspiration for innovation-based and nature-inspired technological research and could be an essential component of sustainable development itself.

1.2.2 Terms and Definitions

Using information and inspiration from living organisms to develop technical artefacts is referred to by a range of different terms like biologically inspired design, biomimetics, biomimicry, bioreplication and bionics. The terms are often used interchangeable, but they are not synonyms, and although similar they each have their special meaning which can be seen in the international ISO standard. The most general term is biologically inspired design which is often used as a synonym for biomimetics but also tends to focus more on the design process. Biomimetics is defined in the ISO-standard as the "interdisciplinary cooperation of biology and technology or other fields of innovation with the goal of solving practical problems through the function analysis of biological systems, their abstraction into models, and the transfer into and application of these models to the solution". Biomimicry denotes "the philosophy and interdisciplinary design approaches taking nature as a model to meet the challenges of sustainable development in a social, environmental, and economic perspective". Bioreplication is the direct reproduction of a biological structure in order to realize at least one type of specific functionality. Bionics constitutes "the technical discipline that seeks to replicate, increase, or replace biological functions by their electronic and/or mechanical equivalents". The bionic was coined by Jack Steele in 1958, being formed as portmanteau from **bio**logy and electro**nic**s. It is the study and application of construction principles and systems in living organisms for the design of (sustainable) technological solutions. Bionics tends to be used in regard to mechanical constructions, whereas biomimetics tends to be used for materials and systems. For example, a building may be referred to as bionic but its surface as biomimetic.Fig.1.3 presents a map of how the different terms relate to one another.

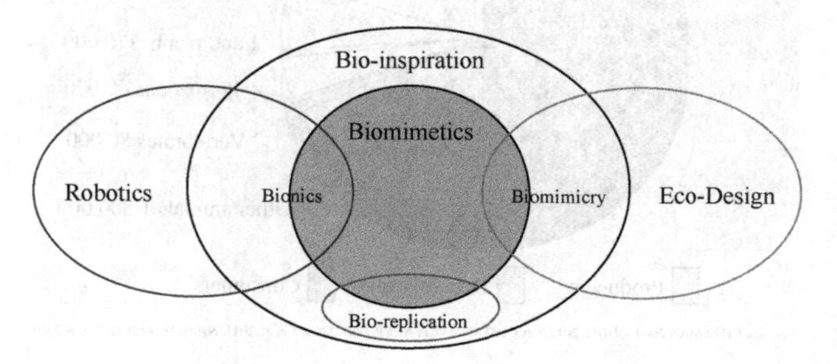

Fig.1.3 Bio-inspiration and linked concepts boundaries map

As the extension of the definition of bionics, it is additionally emphasized that it is important of including aspects of interaction between bio-systems and their environment, which can be applied to economical and management systems as well. The major goal of bionics is not copying of nature or searching for master templates that can be directly transferred into a technical solution. It is even more difficult to define bionic products than bionics itself. Biological species, the living prototypes of bionics, are "sustainable" with regard to their "evolved" life spans.

The terms are often used interchangeably though they have some differences. Therefore, these terms besides bionic may appear in the following content and different chapters.

1.3 Professional Organizations/Institutions

The realization that nature is a continuous optimization process founded a new engineering discipline. Beginning in the 1960s, various schools began to study and to reproduce bionic processes to improve previous optimization solutions. Due to the novel character of the discipline, the availability of approved or established concepts is rather limited, and the best choice is to rely on the example and experience of the German BioKoN network, which might be seen as the most advanced institution for establishing bionic expertise. Among all members of BioKoN, the University of Applied Sciences (Bremen, Germany) is in so far as the most interesting entity, as there, the course "International bachelor's degree in Biomimetics/Bionik" was established already in 2003/2004. For Europe, this is the first institution, which is awarding an official degree in the discipline of bionics.

In 2006, Jilin University set up the second-level doctoral and master's degrees in "Bionic Science and Engineering", which is the first master's and doctoral program in BSE in China. In 2019, the undergraduate program of BSE was established, which strengthens this discipline for system talents training. These official degrees of three systems in BSE was accomplished through the outstanding work of two generations of academicians, Bingcong Chen and Luquan Ren, who also dedicate themselves in the development of the Key Laboratory of Bionic Engineering (Ministry of Education), the International Society of Bionic Engineering, *the Journal of Bionic Engineering*, and the international technology platform of the Bionics International (details as followed).

There are also other research labs, professional societies and networks founded and devoted to bionics. The latest information and development dynamics can be found in these organizations or institutions as introduced below.

1.3.1 The Key Laboratory of Bionic Engineering (Ministry of Education) of Jilin University (KLBE)[1]

The Key Laboratory of Bionic Engineering (Ministry of Education) of Jilin University (KLBE), is a research base of bionic engineering integrating scientific research, technological development, achievement transformation, talent cultivation, international exchanges and discipline construction. With the outstanding achievements in all those aspect, the key laboratory has become an important base of bionic science and engineering with great influence in China and overseas.

The key laboratory adopts the director responsibility system of administration under the guidance of the academic committee, composed of 15 scholars (including 6 academicians of Chinese Academy of Science and Chinese Academy of Engineering). The researchers in KLBE were from nearly 10 different disciplines, working in 4 major research areas, which are: ①Bionic Machinery; ②Bionic Material; ③Bionic Intelligence and Electrical Engineering; ④Bionic Medical Engineering and Healthcare Engineering.

In 2004, *the Journal of Bionic Engineering* was founded. Professor Luquan Ren, the academician of the Chinese Academy of Science, is the editor-in-chief. More than 60 bionic scholars from 19 countries and regions served as the editorial board. In 2010, the International Society of Bionic Engineering (ISBE) was founded by Jilin University in conjunction with bionic scholars from 15 countries, permanently located at Jilin University. The key lab also co-launched the Da Vinci Index of biomimicry, measuring activity in biomimicry and/or bioinspired research and commercial application. In 2019, the Weihai Institute for Bionics-Jilin University was established, which is an important base of research and education belong to the key laboratory.

1.3.2 The International Society of Bionic Engineering (ISBE)[2]

The International Society of Bionic Engineering (ISBE) is an educational, non-profit, non-political organization formed in 2010 to foster the exchange of information on bionic engineering research, development and application.

(1) Mission Statement

The Society is dedicated to the advancement of communication and cooperation among all scholars, and the furtherance of knowledge and education in the field

---

[1] http://bionic.jlu.edu.cn/en/index.htm, retrieved 2022/9/2.

[2] https://www.isbe-online.org/, retrieved 2022/9/2.

of bionic engineering.

(2) Society Main Activities

① Organizing and developing international science communication and collaborations in bionic engineering, encouraging science, and promoting the international development of bionic engineering.

② Popularizing knowledge of bionics and its relevant science and technology, spreading the spirit, idea and methods of the science, and extending bionic technology.

③ Holding international meetings in the field of bionic engineering.

④ Holding meetings of the board of directors.

⑤ Publishing transactions of the society.

⑥ Honoring and encouraging excellent scientists and technologists who make great progress in bionic research and spread activities, and recommend excellent talent.

⑦ Accepting opinions, suggestions and requests from the members, maintaining the legal rights of the members, and developing activities serving the members.

1.3.3 Bionics International (BIN)[1]

The Bionics International is an international technology platform that supports teaching, researching, training, promoting, popularizing, and related activities in the field of bionics with the concept of "comprehensiveness, accuracy, quick, flexibility, and authoritativeness".

Bionics is an emerging interdiscipline that brings together the efficiencies and self-sustaining qualities found in nature to sustainably respond to the complex challenges of this evolving world. In recent years, bionics has developed rapidly and attracted worldwide attention.

The platform serves for the researchers engaged in bionics field and other related enterprises or institutions as well as general public. It is available for users to learn basic knowledge, subscribe bionic news, follow bionic activities, discover scientific resource, reserve lab experiments, establish cooperation relationship, make free discussion, and etc.

The platform is designed in an open mode, and it is a common wish that all of the users could be involved in the construction and development of the platform.

[1] https://www.bionics-online.net/en/, retrieved 2022/9/2.

1.3.4 The Laboratory of Bionics Engineering of the University of Seville (LBE)[1]

The Laboratory of Bionics Engineering of the University of Seville aims to restore lost body functions by delivering precise electrical stimulation. From blindness to gastrointestinal problems, the ultimate goal is to provide medical treatments from a technological perspective. Ranging from electronics to robotics and artificial intelligence, the laboratory of bionic engineering aims to overcome some of the challenges that the field is currently facing.

1.3.5 The Bionics Engineering Lab of the University of Utah (BEL)[2]

They envision a world where everyone can move and live independently. They envision a world where congenital or acquired body differences, trauma, and injury would not prevent people from pursuing their life goals. They envision a world where advanced bionic technologies will enhance the human body restoring, preserving, and augmenting the human movement ability across the lifespan.

To achieve this goal, they focus on the intersection of Robotics, Design, Control, Biomechanics, and Neural Engineering. Their goal is to create new science and develop new technologies empowering the next generation of wearable bionic devices and systems to help people move and live independently, ultimately ending physical disability.

1.3.6 The Center for Extreme Bionics (CEB)[3]

Half of the world's population currently suffers from some form of physical or neurological disability. At some point in our lives, it is all too likely that a family member or friend will be struck by a limiting or incapacitating condition, from dementia, to the loss of a limb, to a debilitating disease such as Parkinson's. Today we acknowledge—and even "accept"—serious physical and mental impairments as inherent to the human condition. But must these conditions be accepted as "normal"? What if, instead, through the invention and deployment of novel technologies, we could control biological processes within the body in order to repair or even eradicate them? What if there were no such things as human disability?

These questions drive the work of faculty members Hugh Herr, Ed Boyden,

[1] https://institucional.us.es/bionicus/, retrieved 2022/9/2.

[2] https://belab.mech.utah.edu/, retrieved 2022/9/2.

[3] https://www.media.mit.edu/groups/center-for-extreme-bionics/overview/, retrieved 2022/9/2.

Canan Dagdeviren, Joe Jacobson, Deblina Sarkar, and Institute Professor Robert Langer, and have led them and the MIT Media Lab to establish the Center for Extreme Bionics. This dynamic new interdisciplinary organization draws on the existing strengths of research in synthetic neurobiology, biomechatronics, and biomaterials, combined with enhanced capabilities for design development and prototyping.

1.3.7 Biomimicry Institute (BI)[1]

The purpose of the Biomimicry Institute is to naturalize biomimicry in the culture by promoting the transfer of ideas, designs, and strategies from biology to sustainable systems design.

We accomplish this by tackling one massive sustainability problem at a time through our Youth Design Challenge and Global Design Challenge, mobilizing tens of thousands of practitioners with the support of the Global Biomimicry Network, and then providing those practitioners with Ask Nature as a tool to begin the solution process. We support entrepreneurs in bringing their nature-inspired designs to market through the Biomimicry Launchpad program and Ray of Hope Prize®, and we anticipate dozens of new biomimetic innovations will result, creating a healthier world for all.

1.3.8 The Institute of Electrical and Electronics Engineers (IEEE)[2]

IEEE, pronounced as "Eye-triple-E", stands for the Institute of Electrical and Electronics Engineers. The association is chartered under this name and it is the full legal name.

However, as the world's largest technical professional organization, IEEE's membership has long been composed of engineers, scientists, and allied professionals. These include computer scientists, software developers, information technology professionals, physicists, medical doctors, and many others in addition to IEEE's electrical and electronics engineering core. For this reason, the organization no longer goes by the full name, except on legal business documents, and is referred to simply as IEEE.

(1) Mission Statement

IEEE's core purpose is to foster technological innovation and excellence for the benefit of humanity.

[1] https://biomimicry.org/, retrieved 2022/9/2.

[2] https://www.ieee.org/, retrieved 2022/9/2.

(2) Vision Statement

IEEE will be essential to the global technical community and to technical professsionals everywhere, and be universally recognized for the contributions of technology and of technical professionals in improving global conditions.

(3) Core Values

① Trust: being a trusted and unbiased source of technical information, and forums, for technical dialog and collaboration.

② Growth and nurturing: encouraging education as a fundamental activity of engineers, scientists, and technologists at all levels and at all times; ensuring a pipeline of students to preserve the profession.

③ Global community building: cultivating active, vibrant, and honest exchange among cross-disciplinary and interdisciplinary global communities of technical professionals.

④ Partnership: promoting a culture of respect for the employees and volunteers, valuing contributions at all levels of the organization, investing in training and development to enhance capabilities, empowering individuals to make a positive difference, and building a membership organization based on a strong volunteer-staff partnership to serve the profession.

⑤ Service to humanity: leveraging science, technology, and engineering to benefit human welfare; promoting public awareness and understanding of the engineering profession.

⑥ Integrity in action: fostering a professional climate in which engineers and scientists continue to be respected for their exemplary ethical behavior and volunteerism.

Key vocabularies

abstraction [æbˈstrækʃn] ***n.*** 抽象；抽象概念；抽象化

arthropod [ˈɑːθrəpɒd] ***n.*** 节肢动物

authentic [ɔːˈθentɪk] ***adj.*** 权威的；可靠的；真正的；真诚的；逼真的

biomimetic [baɪɒmɪˈmetɪk] ***n.*** 仿生的；生物模拟的，拟生态的

biomimicry [ˌbaɪəʊˈmɪmɪkrɪ] ***n.*** 仿生学；生物模拟

bionic [baɪˈɒnɪk] ***adj.*** 仿生学的，利用仿生学的

bionics [baɪˈɒnɪks] ***n.*** 仿生学

charter [ˈtʃɑːtə(r)] ***v.*** 给予……特权

cohabitant [kəʊˈhæbɪtənt] ***n.*** 同居者
colony [ˈkɒləni] ***n.*** 群体；殖民地；聚居人群
fern [fɜːn] ***n.*** 羊齿植物，蕨类植物
gymnosperm [ˈdʒɪmnəspɜːm] ***n.*** 裸子植物
inherent [ɪnˈhɪərənt] ***adj.*** 固有的，内在的；天生的
mutation [mjuːˈteɪʃn] ***n.*** 变化；转变；突变；变异
organism [ˈɔːgənɪzəm] ***n.*** 生物体；有机体；微生物；有机体系，有机组织
portmanteau [pɔːtˈmæntəʊ] ***n.*** 旅行皮箱；混成词. ***adj.*** 综合的；复合式的；多用途的
practitioner [prækˈtɪʃənə(r)] ***n.*** 从业者；习艺者；专门人才；从事者，实践者
predator [ˈpredətə(r)] ***n.*** 捕食者；掠夺者；捕食其他动物的动物，食肉动物
preliminary [prɪˈlɪmɪnəri] ***adj.*** 初步的，初级的；预备的；开端的；序言的
prey [preɪ] ***n.*** 被捕食的动物；捕食；受害者. ***v.*** 捕食；欺凌
prototype [ˈprəʊtətaɪp] ***n.*** 原型，雏形，蓝本
synonym [ˈsɪnənɪm] ***n.*** 同义词
taxonomic [ˌtæksəˈnɒmɪk] ***adj.*** 分类学的
terrestrial [təˈrestriəl] ***adj.*** 陆地的；地球的
vascular [ˈvæskjələ(r)] ***adj.*** 脉管的；血管的；含有血管的；充满活力的
vertebrate [ˈvɜːtɪbrət] ***n.*** 脊椎动物. ***adj.*** 有脊椎的，脊椎动物的
winglet [wɪŋˈlet] ***n.*** 翼梢小翼

Exercises

Ⅰ. Translate the following English into Chinese.

1. Using information and inspiration from living organisms to develop technical artefacts is referred to by a range of different terms like biologically inspired design, biomimetics, biomimicry, bioreplication and bionics.

2. Biodiversity is the inspiration for innovation-based and nature-inspired technological research and could be an essential component of sustainable development itself.

3. The word bionic was coined by Jack Steele, being formed as portmanteau from biology and electronics. It is the study and application of construction principles and systems in living organisms for the design of (sustainable) technological solutions.

4. The International Society of Bionic Engineering (ISBE) is an educational, non-profit, non-political organization formed in 2010 to foster the exchange of information on bionic engineering research, development and application.

5. The realization that nature is a continuous optimization process founded a new engineering discipline. Beginning in the 1960s, various schools began to study and to reproduce bionic processes to improve previous optimization solutions.

Ⅱ. Answer the following questions based on the main text in this chapter.

1. What are the differences and commons among the terms of biologically inspired design, biomimetics, biomimicry, bioreplication and bionics?

2. Why does biodiversity provide the evolved blueprints for bionics?

3. What are the main research fields of the Key Laboratory of Bionic Engineering (Ministry of Education), at Jilin University?

4. What does the ISBE stand for? And what is its main mission?

5. Please search the websites of the professional organizations or institutions and list the latest news and events or meetings interested to you.

Chapter 2

Fundamentals and Principles of Bionics

Bionics as an emerging field at the interface of biology and the world of classical engineering, is gaining more and more acknowledgment and interest from various branches of industry and economy. The already wide spread, but still increasing application of bionic concepts and strategies in product and process development will lead to an increased request for engineers, who have a solid background knowledge in this field, accompanied by a consistent practical experience in the application of the tools and ideas offered by bionics. In order to understand the general approach of bionics, an introduction into the fundamentals and principles of bionics will be presented in this chapter.

2.1 Two Strategic Approaches of Bionics

Starting from the first attempts to construct machines that can fly, researchers and engineers tried to borrow inspirations from nature, thus arriving in some cases at constructions that looked like birds, bats or flying seeds, but not being able to achieve even a minimum percentage of the performance of their natural templates. The final breakthrough in aviation technology is based on the simple recognition, that a functioning technical solution can be achieved by the separation of the parts, which are generating the lift (wings) and the propulsion (engine). Nevertheless, airplane industry is still borrowing ideas from nature, for instance, the introduction of the blended winglet, which is based on bionic ideas and research, derived from studies on the aerodynamics of bird winglets during gliding (Fig.2.1).

The aforementioned example about the development of the blended winglet is, at the same time, a typical example of the top-down approach of the bionic strategy, which means, that starting from a well-defined technical problem, bionics is searching for analogous situations in nature that might provide a solution for that specific problem. The mission of bionics, in this case, is the detailed analysis of the system, which is providing a potential solution, in order to understand how this proposal of nature can be transformed into a technical system with similar properties. The whole process of development and optimization can be seen as a dialog between the worlds of technology and biology, as demonstrated in Fig.2.2. There may be some early results, which appear very similar to the original proposal of nature, but the shape of the final technical solution (like the blended winglet) will in many cases not being traced back to its natural origin.

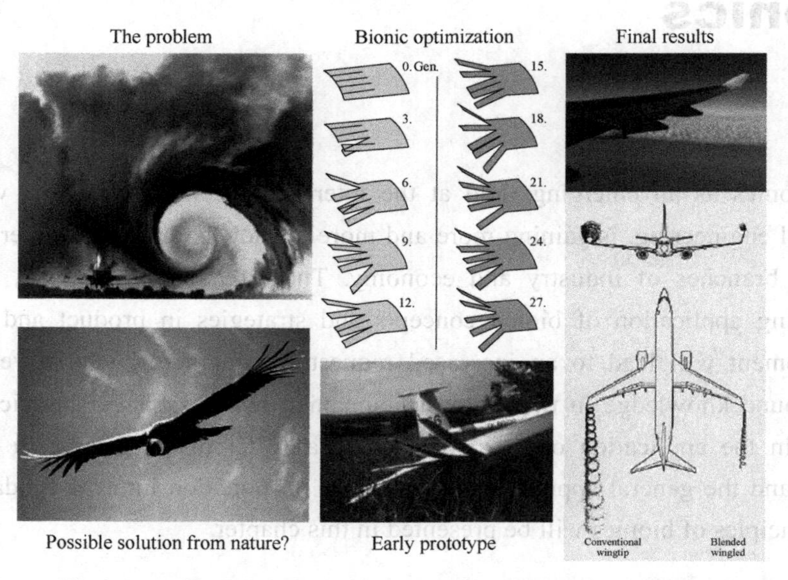

Fig.2.1 Using bionic methods to minimize the energy loss due to vortex generation at the wing tips of airplanes

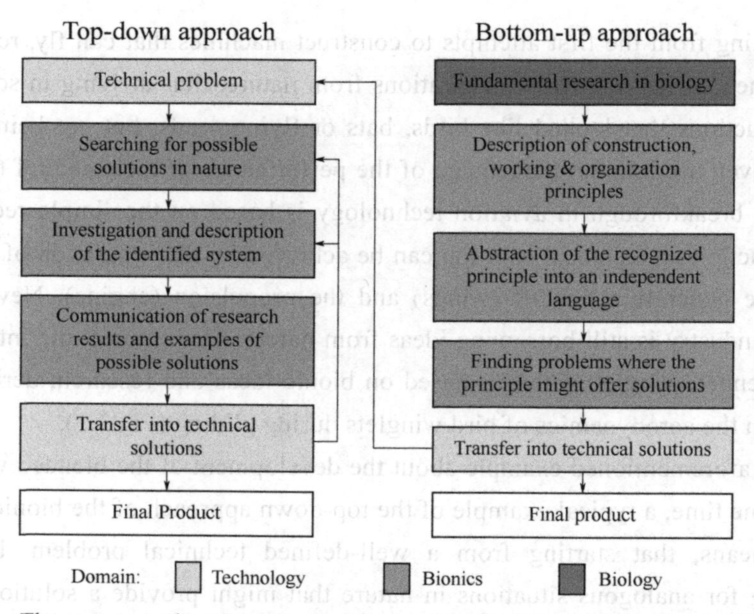

Fig.2.2 The two strategic approaches to integrate bionics into technological development

The complementary strategy, the bottom-up approach of bionics, is not starting with a well-defined problem that has to be solved. It is starting with a (sometimes new) discovery in biological research that is fully described and well understood. The recognized principle then is translated into an abstract, formal and interdisciplinary description of the phenomenon, which is opening the pathways for the transfer into different fields of technology. A well-known example for this approach is the so called "Lotus-effect", which led to many new applications in nano-structured surface design.

Bionics is not to be defined by a specific set of subjects or fields of applications. It is defined by a methodological approach that uses two strategies, in order to transfer inventions of nature into technical solutions. In this way, it is not covering those booming new disciplines like bio-genetic engineering, nano-technology, or bioinformatics, but in many cases, the transfer of ideas and knowledge from biology into the world of engineering, is the starting point for the development of such new disciplines. Therefore, the development of a systematic approach, being in charge with the provision of an interface between fundamental research in biology and innovative engineering, is very important.

2.2 Design Paradigms

Biologically inspired design (BID) is attracting increasing interest since it offers access to a huge biological repository of well proven design principles that can be used for developing new and innovative products. A major obstacle for the wider use of biologically inspired design is the knowledge barrier that exist between the application engineers that have insight into how to design suitable products and the biologists with detailed knowledge and experience in understanding how biological organisms function in their environment. The BID process can therefore be approached using different design paradigms depending on the dominant opportunities, challenges and knowledge characteristics. Design paradigms are typically characterized as either problem-driven, solution-driven, sustainability driven/biomimicry, bioreplication/biohacking or a combination of two or more of them. The design paradigms represent different ways of overcoming the knowledge barrier, and the present paper presents a review of their characterization and application. Each of those paradigms represents a distinct approach for how to address the problem to be solved, how to search and learn from nature and how to do the mimicry. A key issue is the level of abstraction applied in the mimicry, the analysis of problems, and the utilization of biological phenomena. In the following we will describe each of those paradigms in more detail, give typical examples of their application and highlight how the paradigm has been used in various research projects.

Fig.2.3 gives an indication for how much work is done within each of the paradigms. Google Scholar was searched using the search string biomimicry OR biomimetics OR bionik OR bionic OR bioinspired OR "biologically inspired design" resulting in 346,000 publications. By combining the search string with synonyms for each paradigm, the different percentages were found. Surprisingly, problem-driven BID scores relatively high. A possible explanation could be that practitioners within solution-based BID do not use the term "solution-driven" or any of the synonyms and therefore fall into the large "unknown" group.

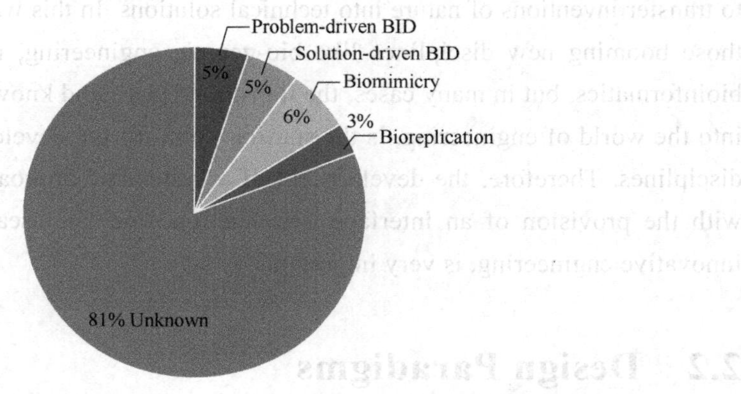

Fig.2.3 Amount of papers in Google Scholar for each of the BID paradigms

2.2.1 Problem-Driven BID

In the problem-driven BID, the starting point is the (engineering) problem for which suitable biological analogies are sought. The term problem-driven BID was coined by the cross disciplinary research group at Georgia Tech, but the paradigm has been followed earlier by several research groups in Canada, Germany, Denmark and United Kingdom. In Germany and Denmark, the term top down biomimetics or top down bionik was used; and in the Netherlands, application-driven bioinspired design was preferred. The paradigm can be regarded as a technology pull where the technology or product determines how to approach and search nature. The AskNature database[1] of biological solutions also applies problem-driven BID in facilitating users with a search engine that can answer the question: how would nature solve this problem?

[1] The Biomimicry Institute has developed the AskNature database (https://asknature.org/), which is a free, open source digital library founded on the principle that truly sustainable solutions will always be inspired by nature. The database presents biological adaptations by function so that engineers, designers and business people alike can look towards nature for design strategies to apply to their own challenges.

Design Model:

Carrying out problem-driven BID requires that a design model is followed. The model specifies the sequence and content of the activities that lead the design team from a design problem to a design solution and could include the following activities: A. problem definition, B. problem framing, C. translation into biological search terms, D. search for biological analogies, E. selection and understanding of how the biological organism works, F. abstraction of key principles, G. design solutions, H. build prototypes and I. evaluate the quality of the developed solution.

The use of a design model can be illustrated with an example which have previously been described in SPIE (the International Society for Optics and Photonics): How to make polymer needles. The problem definition (A) includes an analysis of the present use scenario and the improvement potentials. Today most medical needles for injections are made of metal to assure reliability and easy insertion. However, metal needles are considered dangerous waste after use, as they, unlike the rest of the syringes, need a dedicated waste stream. If the needles could be made of plastic, simpler waste handling would be possible for the combined syringe and needle. But plastic needles have much worse penetration properties than steel equivalents. The problem framing (B) is therefore how to ensure easy insertion of softer polymer needles. The biological search terms (C) could be "skin penetration", "biting" and "stinging". The search (D) finds insects like the stinging mosquito (Fig.2.4 left) and the protection mechanism in cnidarians (Jellyfish), where venomous pointed cells are shot into the threating animal. Understanding the biological analogy (E) can be done through literature study, collaboration with a biologist or a detailed study of the biological organism. For the mosquito, a range of supporting mechanisms like skin stretching, vibration (Fig.2.4 right),

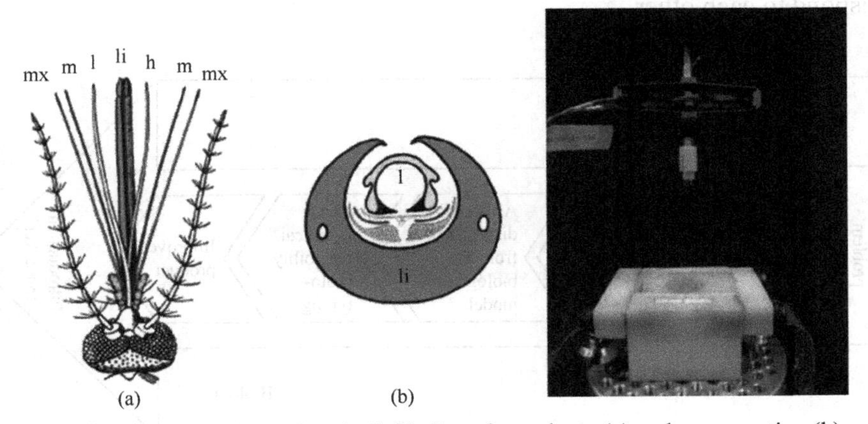

Fig.2.4 The mosquito proboscis (left): Seen from above (a) and cross section (b). Vibration unit (right) mimicking elements of the mosquito stinging mechanism

li—labium, l—labrum, h—hypopharynx, m—mandible, mx—maxilla

impulse and serrations together explain why the fairly flexible and very thin stinging proboscis can penetrate skin. Each of these mechanisms can then be abstracted and described as key principles (F). The principles are used to propose design solutions (G) like a mechanical vibration mechanism that will lower the penetration force for a needle and hence reduce the requirements to its stiffness. This can be verified empirically with a prototype (H) and an evaluation (G) where the proposed solution is compared with similar solutions.

There exist a number of different design models covering all or some of the activities described above. The ISO-standard mentions a sequential model with 5 steps as shown in Fig.2.5. The focus is on the search and understanding of biological analogies and on the abstraction from the biological model. The initial phase where the problem is analyzed and formulated is not treated in the standard. The Biomimicry Institute applies a design model that they call the design spiral shown in Fig.2.6(a). It has 6 steps which cover all activities from the initial problem analysis (identify) to the final evaluation of the developed product. The shape of a spiral communicates that the model can be repeated in part or whole until a satisfactory result is achieved. Distinct contributions include the biomimicry taxonomy for problem framing, the AskNature database for search and life's principles for sustainability evaluation. The design model used by the Georgia Tech BID group also comprises 6 steps as shown in Fig.2.6(b). The steps in the model remind of the step in the design spiral. The major difference is the final evaluation step which is not directly mentioned in the Georgia Tech model. The single steps in those 3 design models are compared in Table 2.1 which shows how the models correspond to each other.

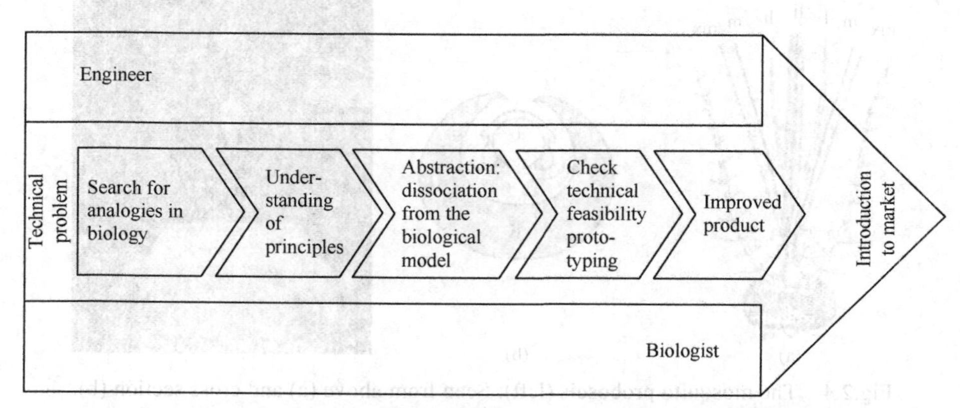

Fig.2.5 Typical technology pull BID development process as described by ISO

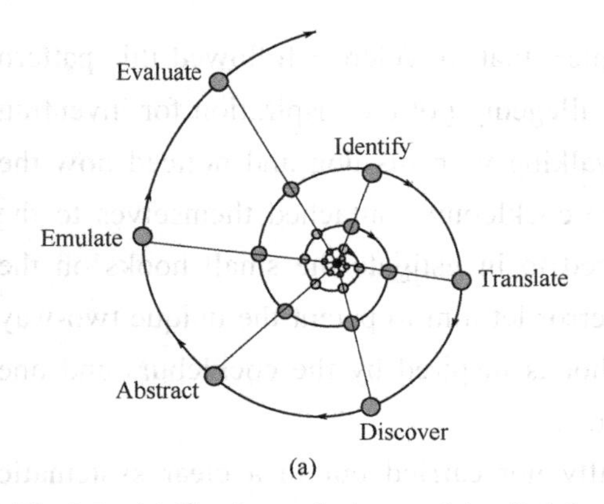

(a)

Step 1: problem definition

Step 2: reframe the problem

Step 3: biological solution search

Step 4: define the biological solution

Step 5: principle extraction

Step 6: principle application

(b)

Fig.2.6 (a) The 6 step design model called the design spiral used by the Biomimicry Institute, (b) The 6 steps in the Georgia Tech BID design model

Table 2.1 Comparison of 3 design models for problem-driven BID①

	A problem	B frame	C translate	D search	E select & understand	F abstraction	G design	H build	I evaluate
ISO model	0. Technical problem			1. Search	2. Understand principles	3. Abstraction from biological model			4. Check technical feasibility
Biomimicry Institute design spiral	1. Identify		2. Translate, Biomimicry taxonomy	3. Discover, AskNature		4. Abstract	5. Emulate		6. Evaluate, Life's principles
Georgia Tech model	1. Problem definition, 4box	2. Reframe the problem, T-chart		3. Search	4. Define the biological solution	5. Principle extraction	6. Principle application		

① Light grey fields mark activities covered, darker grey fields are where the model has a distinct contribution.

2.2.2 Solution-Driven BID

In the section above, we learned how designers can systematically search for biologically inspired answers to clearly defined engineering problems via the problem- driven or top-down approach. However, most biomimetics studies have so far been conducted by the alternative solution-driven or bottom-up approach. This method is less systematic and typically occurs as a result of serendipitous discoveries of fascinating biological processes or mechanisms.

Solution-driven BID takes the fascination of interesting and remarkable biological phenomena as a starting point and subsequently analyses these to identify key principles and then search for suitable applications. The paradigm has been phrased solution-driven BID, bottom up biomimetics, inspiration-driven biologically inspired design and biology push.

The most famous biomimetic example—that of Velcro—followed this pattern as the Swiss engineer George de Mestrel allegedly got the inspiration for inventing Velcro in the 1940s, when he was out walking with his dog and noticed how the seed chambers of the burdock plant—the cockleburs—attached themselves to the fur of his dog. Upon his return, he started to investigate the small hooks on the cockleburs, which after a lot of trial and error let him to patent the unique two-way fasteners consisting of one side of tiny hooks inspired by the cockleburs and one side of soft loops inspired by the dog's fur.

As mentioned, this process is usually not carried out in a clear systematic manner, but some researchers nonetheless developed a framework for solution-driven BID consisting of the following 7 steps:

Step 1: Biological Solution Identification—This is where the biological process or mechanism is encountered or discovered (for example, cockleburs forming strong attachments to dog fur).

Step 2: Define the Biological Solution—Here the mechanisms are described and explained (tiny hooks make the cockleburs attach to the fur).

Step 3: Principle Extraction—The core principle is extracted from the biological process or mechanisms (multiple stiff hooks adhere to multiple looped hairs).

Step 4: Reframing the Solution—How might the biological mechanism or process be redesigned to be applied to human technology (multiple mini hooks can form strong attachment to multiple mini loops)?

Step 5: Problem Search—Exploration of which problems in our technology this principle can be the solution to. This may sometimes involve developing novel problems—i.e. new applications or novel technology (the hooks and loops principle can dynamically be used to attach and detach).

Step 6: Problem Definition—Narrowing down the problems identified in step 5 to extract a clearly defined problem or application (the hook and loop dynamic fastener principle could be used in fabrics that need to attach and detach rapidly such as shoes, jackets, etc.).

Step 7: Principle Application—Apply the developed principle to prototypes and final products (apply the Velcro hooks and loops to clothes and compare to traditional methods).

In most examples, where the solution-driven BID approach has been used (usually without this being explicitly acknowledged), steps 1-3 and sometimes 4 and 5 have been carried out by biologists, who are experts on the given organisms and by accident or due to scientific curiosity discover the novel mechanisms.

In the following, we give examples of how this approach has been utilized (again without implying that the scientists were doing this in a systematic way or consciously using the steps outlined above) in three of the most famous biomimetics examples and include some of the challenges encountered in successfully translating the biological solutions.

2.2.2.1 The Cleaning Effect of Lotus Leaves

The Lotus effect is the mechanism by which small epicuticular wax crystals on the leaf surface of some plants, including the sacred lotus plant, allow the plants to self-clean by giving water a stronger adhesion to dirt particles than to the leaf surface as shown in Fig.2.7. The mechanism was discovered accidentally when botanists, who were investigating the surfaces of leaves from a large number of species, observed that smooth leaves were much dirtier than rough leaves (step 1). From these initial observations, the exact nature of the roughness and its hydrophobic effect was quantified, and the principles of their function were extracted under the name of the Lotus effect (step 2 and 3). These steps were all completed by the botanists, led by Willem Barthlott, who discovered the mechanism, and they further suggested possible applications in self-cleaning surfaces and paint (step 4 and 5). Barthlott patented the Lotus effect in 1995, and the company Sto SE &Co acquired the patent and started producing the paint Lotusan (step 6 and 7), which is now commercially available and has demonstrated advantages including being more cost effective compared to traditionally paints. Thus, the Lotus effect is a clear example of a successful application of solution-driven BID.

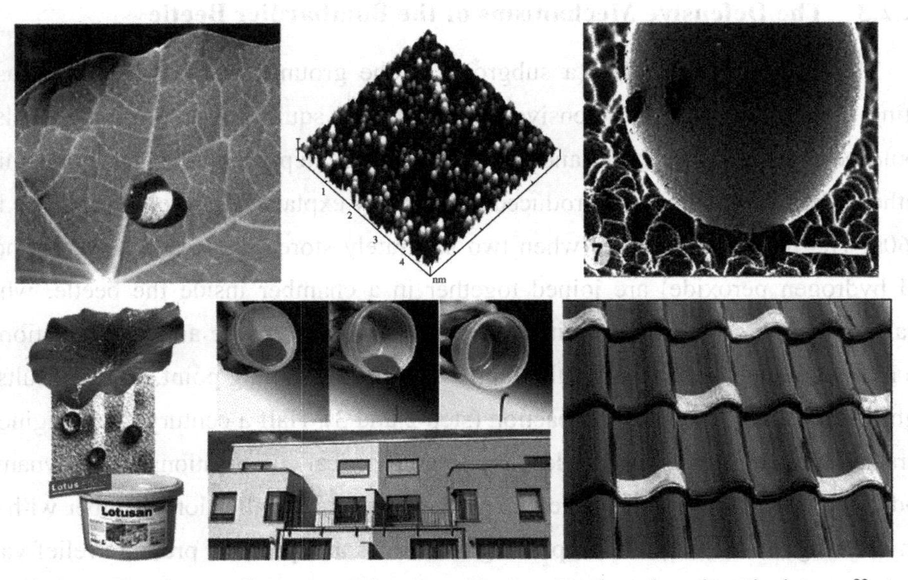

Fig.2.7 Top: Lotus-effect at work in nature. Bottom: Products based on the lotus-effect (Paint, Self-cleaning food container, Self-cleaning roof)

2.2.2.2 The Material Properties of Spider Silk

It has long been recognized that spider silk is a remarkable biomaterial allowing, among other things, spiders to catch fast flying and large prey with an ultralight weight web. It is particularly the frame threads (also used in the safety threads that spiders trail behind them when moving) that due to their combination of high strength and high elasticity, has an exceptionally high toughness, which is necessary in order to absorb the high kinetic energy of the prey (step 1). The biochemical structure of the silk is also fairly well understood, and it is the composite nature of the silk consisting of amorphous flexible chains and stiff polyalanine arranged into hydrogen-bonded beta-pleated sheets (step 2 and partly 3). However, the process of spinning occurring inside the spider, where the silk changes from a liquid phase in the glands and ducts to the solid phase extruded by the spinnerets via a complicated mechanical and chemical process, is still not fully understood (step 3 not fully completed). Thus, while biomimetic applications including in protective clothes and crash helmets have been identified (step 4 and 5), and the biochemical structure is known and the liquid silk proteins can be produced by genetically modified bacteria, the biomimetic spinning process has not yet been mastered and artificial silk is therefore currently much inferior to natural silk. However, while commercial products relying on spider silks' unique material properties are a long way off (step 6 and 7 not achieved), the biocompatibility and organic breakdown of artificial or reconstituted silk makes it potentially very useful as suture threads in surgery and for other biomedical applications.

2.2.2.3 The Defensive Mechanisms of the Bombardier Beetle

The bombardier beetles (a subgroup of the ground beetles) display a fascinating and highly efficient defensive behavior. They squirt hot noxious chemicals at would-be predators in a rapid and directed fashion (step 1). The exact biochemical method whereby this liquid is produced was already explained by biologists in the late 1960s. The liquid is produced when two separately stored chemicals (hydroquinone and hydrogen peroxide) are joined together in a chamber inside the beetle, where catalysts facilitate the decomposition of the hydrogen peroxide and the oxidation of the hydroquinone. This heats up the liquid close to the boiling point, which results in high pressure and an explosive reaction (step 2 and 3). Half a century later, engineers extracted the main principles by developing a numerical computational fluid dynamics model and using it to build a scaled up prototype of the combustion chamber with the aim of using the technique to improve gas turbines and produce pressure relief valve systems that can deliver a periodic pulsed spray (step 4 and 5). The company Swedish Biomimetics 3000 is attempting to utilize this technology in μMIST® to develop the ability to spray highly viscous formulations, without requiring flammable and environmentally damaging propellants (step 6). However, to date no commercially available products have been developed (step 7 not fully achieved).

2.2.2.4 Status of Solution-driven BID

Thus, in conclusion, solution-driven BID has generated a large number of novel biomimetic applications, but as illustrated with the examples selected above, few have as of yet resulted in commercially successful products (only Velcro and Lotusan can be classified as successes). The main challenges are due to insufficient understanding of the biological mechanisms (spider silk), misapplying problems with simplifying and replicating the process (spider silk and defensive mechanism in bombardier beetles). It is also worth noting that in all of the examples highlighted here, the first 3-4 steps (and for the Lotus effect, the first 6 steps) were completed by biologists before engineers and physical scientists took over, illustrating the need for including both biologists and engineers in the solution-driven BID process.

2.2.3 Biomimicry

The term "biomimicry" scopes the technical transformation of biological forms, processes, patterns, and systems into artefacts. The connection of biomimicry to sustainability is inherent, as the evolutionary process of nature creates resource efficient results which should support the solution of ecological challenges caused by mankind. Cohen and Reich generated an overview of terminology of different biomimetic design directions. They argue that the sustainability driven BID as a paradigm cannot be separated from the paradigms described above. It occurs as problem-driven as well as solution-driven approaches. The difference is the explicit focus on sustainability, especially the environmental sustainability. Therefore, this section will not only consider the term biomimicry, but also look at bio-inspiration, biomimetics and bionics, if their application supports at least one of the criteria of sustainability.

2.2.3.1 Sustainability

Traditionally sustainability is described including three dimensions: the economic, the social and the environmental dimensions. United Nations has recently further concretized these dimensions into sustainability goals. The focus of sustainability-driven BID is in many cases the environmental dimension, but especially the evaluation of the results from a sustainability-driven BID process is not meaningful without the economic dimension. In the following, we will mainly be considering the environmental dimension of sustainability.

It is often assumed that a BID product or system is automatically sustainable, but that is not always the case. For example, a light weight product inspired by nature might save economic and environmental resources in building it, but the geometric shape might be so complex, that the production cost exceeds the saved material costs. Another example is poisonous animals and plants, which are the result of a natural production process but nevertheless are harmful to humans and other natural beings. The poisonous organisms serve their purpose within their ecological niche, but the poisonous element cannot be taken out of context claiming sustainability. The descriptive, normative and emotional aspects of bio-derived developments are taken into consideration when looking at sustainability. The descriptive aspect aims at facilitating knowledge transfer between similar BID applications, and proposes a classification into classes for this purpose. The normative aspect aims at evaluating to what extend a proposed solution is sustainable, while the emotional aspect deals with the comprehension of the product and the biological analogy. The Biomimicry Institute proposes to use the Life's Principles to evaluate the sustainability of a product.

For the evaluation of environmental sustainability, quantitative methods for life cycle assessment (LCA) are widely used. The LCA-tools calculate the impact of a product on the environment and translate it into indicators such as CO_2-equivalents. The LCA-result is not an absolute measurement of sustainability, but gives a relative number that can be used to compare new solutions to existing ones.

2.2.3.2 Biomimicry Levels

The Biomimicry Institute defines 3 levels of biomimicry having increased complexity. The first level of biomimicry concerns the mimicking of natural forms, such as the imitation of frayed owl wing edges to achieve silent flight. The second level scopes the transfer of natural processes into technical solutions, for example, the way green chemistry is inspired by self-assembly of owl feathers at room temperature. The third level has the highest requirement as it covers the mimicking of natural ecosystems, also named as deep, holistic or eco biomimicry. It implicates the integration of all elements and the relationship between the elements in a more or less closed ecosystem. An example is the consideration of the owl feather as part of a forest and a sustaining biosphere. To achieve sustainable solutions, level 3 biomimicry should be aimed for. Solutions found at the 2 lower levels need not being sustainable.

An engineering counterpart to the biomimicry levels is classified into a number of levels with increasing complexity. One is the distinction between the organism level, the behavior level, and the systems level, as an approach which stems from mechanical engineering design. We have used five levels to classify 5 examples of biomimicry: the system level on the top followed by products/devices, components/ modules, elements/ parts and materials. Table 2.2 places the example at the different levels and shows their relation to the biomimicry levels. The technical examples seem to indicate the following tendency: using natural forms are mainly to be found on the technical site at the element and component level. Natural processes (as second level) are implemented from the material up to products and devices, whereas mimicking ecosystems are mostly applied to the systems and product level.

In megacities, there is increasing demand for climate regulation and purify-cation of the air. This has increased the interest in natural mechanisms for air filtration, the increased plantation of trees and the concept of an urban green infrastructure. In extension to the contribution of parks and tree lined alleys, there is an increasing relevance of designing and implementing items of urban greenery to green roofs and facades. However, as urban trees grow significantly faster than their rural conspecifics, they manifest a more rapid ageing and thus shortened lifetime which induces the need for new concepts integrating plants into megacities to ensure their contribution to the urban ecosystem. The example illustrates that it is not enough to mimic (or in this case) the natural organism, but the ecological system in wood has to be considered.

Table 2.2 Five examples of biomimicry classified using the biomimicry levels and a traditional engineering hierarchy of technical systems

Hierarchy of technical artefacts	Application	Mimicking	Biomimicry levels
Systems	Purification of air in megacities	CO_2 uptake and renewal of plants	3 Eco-systems
Products, devices	Ventilation in buildings	African termite mound	
Components, modules	Blades of fans and turbines, facades	Bumps on humpback whale fins	1 Natural Forms
Elements, parts	Infrared sensor for fire-detection in buildings	Pyrophilous beetle	2 Natural Process
Material	Impact protection	Spider silk	

Key vocabularies

abstract [ˈæbstrækt] ***n.*** 摘要；抽象的概念. ***v.*** 抽象化；提取，使分离

allegedly [əˈledʒɪdlɪ] ***adv.*** 据说，据称

amorphous [əˈmɔːfəs] ***adj.*** 无定形的；无组织的；非晶形的

as of yet 到目前为止
biocompatibility [bi:əʊkəmpætəˈbɪlɪtɪ] **n.** 生物相容性；生物适合性
biosphere [ˈbaɪəʊsfɪə(r)] **n.**生物圈；生物界；生物层
bombardier beetle **n.** 放屁虫
burdock [ˈbɜ:dɒk] **n.** 牛蒡，牛蒡属
chamber [ˈtʃeɪmbə(r)] **n.** 房间；腔，房室
cnidarian **n.** 刺胞动物
cockleburs [ˈkɒk(ə)lbɜ:] **n.**苍耳属植物
concretize [ˈkɒnkri:taɪz] **vt.** 使具体化，使有形化
conspecific [ˌkɒnspɪˈsɪfɪk] **adj.** 同种的
counterpart [ˈkaʊntəpɑ:t] **n.** 对应的人或事物
epicuticular [ˌepɪkju:ˈtɪkjʊlə] wax **n.** 表皮蜡，角质层蜡
equivalent [ɪˈkwɪvələnt] **adj.** 等同的，等效的. **n.** 对等的人（或事物）；当量
explicitly [ɪkˈsplɪsɪtlɪ] **adv.** 清楚明确地，详述地；坦率地；不隐晦地
facade [fəˈsɑ:d] **n.** 建筑物的正面
facilitate [fəˈsɪlɪteɪt] **v.** 使更容易，使便利；促进，推动
fray [freɪ] **v.** 磨损，磨松
gland [glænd] **n.** 腺
holistic [həˈlɪstɪk] **adj.**全盘的，整体的；功能整体性的
hydrophobic [ˌhaɪdrəˈfəʊbɪk] **adj.** 疏水的
hydroquinone [haɪdrəkwɪˈnəʊn] **n.**对苯二酚；氢醌
hypopharynx [ˌhaɪpəˈfærɪŋks] **n.** 下咽部，喉咽（部）
impulse [ˈɪmpʌls] **n.** 脉冲；搏动；推动力. **v.** 推动
jellyfish [ˈdʒelifɪʃ] **n.** 水母，海蜇
labium [ˈleɪbɪəm] **n.** 唇；阴唇；下唇瓣
labrum [ˈleɪbrəm] **n.**（尤指昆虫的）上唇
loop [lu:p] **n.** 环形；环状物；回线，回路；环线. **v.** 使成环，使绕成圈
mandible [ˈmændɪbl] **n.**下颚，下颚骨；下牙床
manifest [ˈmænɪfest] **v.** 显示，表明；（病症）显现
maxilla [mækˈsɪlə] **n.** 上颌骨；（昆虫的）下颚；（甲壳类的）小颚
mimicry [ˈmɪmɪkri] **n.** 模仿，模拟；模仿的技巧；（动物等）拟态伪装
mosquito [məˈski:təʊ] **n.** 蚊子
niche [ni:ʃ] **n.** 生态位（一个生物所占生境的最小单位）
normative [ˈnɔ:mətɪv] **adj.** 标准的，规范的
noxious [ˈnɒkʃəs] **adj.** 有害的；有毒的；败坏道德的；讨厌的
paradigm [ˈpærədaɪm] **n.**范例，样式，模范；样板，范式；词形变化表
penetration [ˌpenɪˈtreɪʃn] **n.** 渗透；穿透，进入
pleated [pli:t] **n.** 褶；褶状物. **v.** 使……打褶

polyalanine ***n.*** 聚丙氨酸
polymer [ˈpɒlɪmə(r)] ***n.*** 多聚物；聚合物
proboscis [prəˈbɒsɪs] ***n.***（象等的）长鼻；（昆虫等的）喙；探针
propellant [prəˈpelənt] ***n.*** 推进物；推进燃料；发射火药. ***adj.*** 推进的
propulsion [prəˈpʌlʃn] ***n.*** 推进；推进力
repository [rɪˈpɒzətri] ***n.*** 仓库；贮藏室；博物馆；数据库
sacred [ˈseɪkrɪd] ***adj.*** 神圣的；宗教的；受尊重的，不可侵犯的，不容干涉的
scenario [səˈnɑːriəʊ] ***n.*** 设想；可能发生的情况；剧情梗概；场景
scope [skəʊp] ***n.*** 范围，领域；视野，眼界. ***v.*** 评估，调查，确定
serendipitous [ˌserənˈdɪpətəs] ***adj.*** 偶然发现的，侥幸的，凑巧的
serration [seˈreɪʃn] ***n.*** 锯齿状；锯齿状突起
spinneret [ˈspɪnəˌret] ***n.*** 喷丝头；吐丝器
squirt [skwɜːt] ***v.***（使）（液体）喷射，喷涌
stiffness [ˈstɪfnəs] ***n.*** 僵硬；生硬；强直；顽固；挺度，劲度，刚性，坚硬度
suture [ˈsuːtʃə(r)] ***n.*** 缝合，缝伤口；缝线. ***vt.*** 缝合伤口
syringe [sɪˈrɪndʒ] ***n.*** 注射器，注射筒；洗涤器. ***vt.*** 注射
thread [θred] ***n.*** 线；细线，线状物
velcro [ˈvelkrəʊ] ***n.*** 魔术贴，尼龙搭扣；维克罗（尼龙粘扣商标名）
venomous [ˈvenəməs] ***adj.***（蛇等）分泌毒液的；有毒的
vibration [vaɪˈbreɪʃn] ***n.*** 摆动；震动；（偏离平衡位置的）一次性往复振动
viscous [ˈvɪskəs] ***adj.*** 黏的；黏性的；半流体的；黏滞的
vortex [ˈvɔːteks] ***n.*** 涡流；涡旋；旋风

Exercises

Ⅰ. Translate the following English into Chinese.

1. The Lotus effect is the mechanism by which small epicuticular wax crystals on the leaf surface of some plants, including the sacred lotus plant, allow the plants to self-clean by giving water a stronger adhesion to dirt particles than to the leaf surface.

2. The biochemical structure of the silk is also fairly well understood, and it is the composite nature of the silk consisting of amorphous flexible chains and stiff polyalanine arranged into hydrogen-bonded beta-pleated sheets.

3. Half a century later, engineers extracted the main principles by developing a numerical computational fluid dynamics model and using it to build a scaled up prototype of the combustion chamber with the aim of using the technique to improve gas turbines and produce pressure relief valve systems that can deliver a

periodic pulsed spray.

4. The connection of biomimicry to sustainability is inherent, as the evolutionary process of nature creates resource efficient results which should support the solution of ecological challenges caused by mankind.

5. An engineering counterpart to the biomimicry levels is classified into a number of levels with increasing complexity. One is the distinction between the organism level, the behavior level, and the systems level, as an approach which stems from mechanical engineering design.

Ⅱ. Answer the following questions based on the main text in this chapter.

1. What are the two major approaches to integrate bionics into technological development?

2. What are the general activities leading the design team from a design problem to a design solution?

3. What are the major steps in the sequential model of biologically inspired design process for top-down approach as described by ISO-standard?

4. What are the general steps for solution-driven approach developed by some researchers?

5. What are the connections between biomimicry and sustainability?

Ⅲ. Writing.

Search a journal paper related to the bioinspired design using the problem-driven approach, from the database of Web of Science, and summarize the general steps following one of the design models mentioned in this chapter.

Chapter 3

Enhance the Success of Biomimetic Programs

Biomimetics can be performed on a bottom-up or top-down basis. Both approaches require a clear identification and isolation of the working principles from the biological model, due to the fundamental differences between living and engineered systems. There are compelling reasons for modelling engineering solutions on natural processes. Since it lacks foresight, evolution by natural selection can arrive at solutions unlikely to be devised by human ingenuity. Moreover, step-by-step modifications of traits over thousands or millions of generations allow for the assembly of complex hierarchical structures that comply with environmental demands. The traits of an organism have redundancy, and hence robustness, because they must respond to multiple competing requirements. Organisms can also self-repair as a corollary of their ability to change plastically during development.

One could point out success stories, where basic biological research and biomimetics have had significant economic and scientific impacts. For example, the discovery of the Lotus-effect pushed the boundaries in water-repellent and self-cleaning surfaces, and led to multiple commercially successful products. Aircraft engineering has always drawn from the study of bird wings, while advances in biotechnology, bioengineering, biomedical engineering and pharmacy are founded on the mimicry of biological substances and processes. On the other hand, for many ambitious and well-known long-term programs, such as the fabrication of a biomimetic spider silk, significant breakthroughs are still pending. With continuing advances building on several years of practical experience, it is now the time for a critical assessment of the efficiency of biomimetics.

Here we broach the question: which factors and strategies have led to success or hampered the advancement of biomimetic programs? First, we briefly review the discrepancies between biological systems and engineering designs that researchers must acknowledge when using biomimetics. Then, we examine two long term biomimetic research programs in an attempt to identify common problems and pathways to success. Finally, we suggest some more directed and defined processes to enhance the success of biomimetic projects.

3.1 Biomimetic Approaches to Engineering Designs May be Suboptimal

In effect, biomimetics assumes that evolution by natural selection is a series of natural experiments that have optimized a design and rendered any suboptimal alternatives extinct. Biologists question this assumption, highlighting the difficulties that may arise because of systematic differences between evolution by natural selection and engineering procedures. In contrast to engineering, evolution by natural selection:

① has a constrained starting-point, namely the traits of the organism as they exist at any moment in time, whereas engineers potentially arrive at a solution by choosing any convenient starting point.

② acts on organisms that draw on materials available in the local environment, whereas engineers can utilize materials taken from any environment.

③ responds to contemporary rather than future requirements, and hence has no foresight. In contrast, engineering processes can assess alternative solutions based on criteria such as sustainability.

④ responds to multiple competing requirements imposed by the environment. The traits of an organism therefore almost always represent some degree of trade-off between a multitude of functions, rather than being optimized for a single function.

The awareness of these limits is an important pre-condition for successful biomimetics, and ignorance of these distinctions is a common pitfall in the process of working principle extraction.

In the following, we take a closer look at two distinct cases and investigate how the process of identification of working principles and their transfer onto a technical model has been realized.

3.2 Cases of Evaluation of Bionic Design

3.2.1 Designing Reversible Adhesives Based on Gecko Toe Pads

Most physical and biological interactions between an organism and its environment take place on surfaces. Accordingly, biological surfaces perform a variety of tasks, and were the first biological system to capture the interest of physicists and engineers seeking new ways to push boundaries and create novel materials. One prominent example is the dry adhesive toe pad system of gecko lizards. Gecko toe pads are more strongly and reversibly adherent to most surfaces than synthetic adhesives. There is therefore immense interest in developing adhesives that mimic the properties of gecko toes.

3.2.1.1 The Working Principles of Gecko Toe Pad Adhesion

In order to design gecko-inspired adhesives from synthetic polymers, it has been essential that engineers understand the basic physical mechanisms acting in gecko toe pad surfaces. Researchers have used techniques such as scanning electron microscopy to examine the microscopic hair-like keratinous protuberances called setae on the gecko toe. Geckos lacking setae cannot climb smooth surfaces. At their tips, the setae subdivide into finer nano-branches with flattened endings called spatulae. It is believed that these structures are so pliable that they can get exceptionally close to a naturally irregular substrate. Adhesion tests on synthetic surfaces have shown that the toe pads stick to surfaces via van-der-Waals forces between the spatulae and the substrate. The scale of the interaction is so small that even tiny dust particles can impede the adhesive mechanism. The gecko system is nevertheless efficient at self-cleaning, and full adhesive capacity is recovered after just a few attachment-detachment cycles. Rapid switchability between high and low adhesion was attributed to the anisotropy of the setae, which cyclically align, misalign and re-align with the surface by shear forces. These principles have duly been identified for the development of gecko-inspired adhesives.

3.2.1.2 Designs of Gecko-inspired Adhesives

Projects aimed at developing gecko-inspired adhesives have almost without exception focused on the setae as the primary structure providing the desired functionality. As there were initially no fabrication methods that could produce such fine scaled structures, simplified derivatives were used. Some of these emphasized the fibrillary character of the setae, others emphasized the spatula-shaped contact elements, while others emphasized a tilting of the fibrous arrays, structural anisotropy, or the multiple hierarchy (i.e. “hairs on hairs”) of the structures (Fig.3.1).

Choice of the key feature appears to have been most commonly determined by the fabrication methods available and the properties of the usable materials. For instance, fibrillary characters were apparently chosen not on the basis of functional analysis but because fiber arrays are considerably easier to produce than complex spatulate structures. It has also emerged that competing properties in any fibrous adhesive must be balanced to function adequately. For example, if the fibers are too stiff and thick, they are not flexible enough for effective adhesion; but when they are too soft and thin, they break under load or stick to each other rather than to the substrate. Arguably, excessive branching and minute spatula-shaped contacts are important features that facilitate the high efficacy of the gecko adhesive system, because they enable extreme compliance with a relatively stiff building material (keratin), which is simultaneously durable and wear-resistant under excessive loads. This highlights the danger of focusing too much on a single feature of a living system in a biomimetic approach where multi-functionality is the key to the design goal.

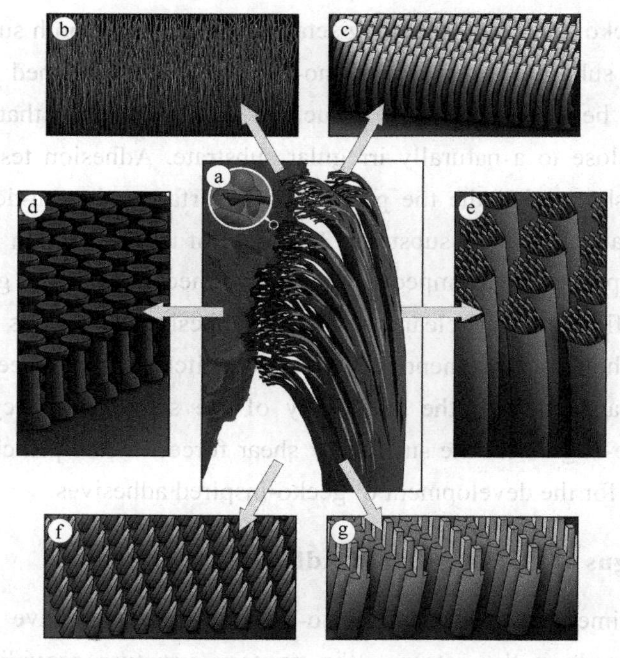

Fig.3.1 Pathways of extracting the essential feature(s) in gecko-inspired adhesives. The adhesive hairs (setae) of geckos [(a) schematic view of a single seta tip in contact with a plant surface] split into numerous branches with nano-scale flattened tips, so-called spatulae [(a) inset]. This structure combines stability with a high degree of conformability, such that spatulae can closely contact the substrate surface so short range inter-molecular forces can be utilized. Material engineers attempt to divide this complicated structure into simplified models to mimic the working principles. Thereby single structural features are often used in isolation: e.g. high aspect ratio nano-fibers (b); tilted micro-columns (c); micro-columns with discoidal tips (d); bi-hierarchical fibers with flattened tips (e); wedge-shaped elements with shear-sensitive adhesion (f); tilted micro-columns with flexible tips (g). Notably, none of these materials exhibits adhesive properties similar to gecko toes

Some studies have circumvented this problem by applying a broader view of natural model systems. Instead of focusing exclusively on the gecko model, broad comparative studies of insects and spiders have revealed a diversity of functional structures for adhesion. Some of these might be easier to transfer into manufactured products, or might more closely match a particular designer's goal. Notably, spatulate structures such as those found on gecko toe pads, are highly beneficial for rapid attachment-detachment cycles, but the design goal of most gecko-inspired adhesives is longer term attachment. Surface features that have evolved for strong but not dynamic attachment exhibit contact elements with entirely different shapes. This principle was used in the development of one of the few commercially developed dry adhesive tapes derived from biomimetic procedures (Gecko® Nanoplast®, Gottlieb Binder GmbH & CoKG, Holzgerlingen, Germany). A comparative study of gecko toe pads proposed that focusing on setae as the fundamental feature might be ineffectual because their properties cannot be scaled up. These authors emphasized the mechanical properties of the underlying material and developed a fabric-polymer blend (Geckskin™, University of Massachusetts, Amherst, MA, USA) which mimics the anisotropic compliancy of a gecko toe but does not mimic any of the fibrillary features.

3.2.1.3 Evaluation

The difficulties associated with developing adhesives inspired by gecko toe pads highlight several problems with biomimetic design approaches. One is an apparent discrepancy between the biological function and the intended, often not clearly defined, application. Gecko-inspired adhesives ought to stick instantly on various surfaces and should be removable without leaving marks. Many authors claim that their developments would perform as well as or even better than the natural model. However, such claims are usually made upon evaluation of only a single characteristic (e.g. perpendicular pull-off strength, dynamic friction, or self-cleaning capacity) and/or by using a particular testing method, substrate surface, or sample type. Rigorous testing of performance relative to conventional synthetic adhesives is rare. The efficacy of gecko-inspired adhesives is particularly difficult to evaluate since the pull-off forces are usually measured with reference to the direct contact between the micro-hair tips with a substrate rather than the effective adhesive area. This means that it is often unclear whether a non-structured flat sample of the same material produces similar adhesive and friction forces as the bioinspired one. A further problem is the restrictive and highly unnatural functional analyses performed for the gecko adhesive system. For instance, although geckos stick to any surface regardless of how smooth, rough, dirty, or wet, a feat which cannot be matched by any synthetic adhesive tapes, most tests are performed on smooth polar artificial surfaces. Overall, it appears that poor understanding of the natural model has been the main reason for the long-term trial-and-error process observed in the development of gecko-inspired adhesives.

3.2.2 Development of High Performance Materials Based on Spider Dragline Silk

Spider dragline silk is an exceptional material with a unique combination of high tensile strength and extensibility. Its toughness exceeds that of most natural and synthetic materials, including Kevlar®. Moreover, it is produced within an aqueous solution at room temperature and is highly biocompatible. The production of artificial fibers that mimic the properties of dragline silk is therefore sought-after. Potential applications include novel light-weight, high-performance materials (e.g. ropes, protective clothing) and functional bio-composites for tissue engineering. Harvesting silk from spiders, as opposed to silkworms, is commercially unviable as spiders require vast amounts of space for their webs, tend to cannibalize each other, and do not readily produce large quantities of silk. Genetic engineering procedures utilizing biomimetic spinning methods appear to be the best option for the large scale production of high performance spider silks.

3.2.2.1 Structure-property Relationship in Natural Spider Silk

Detailed studies have established the links between the expression of certain spider silk genes and the proteins (spidroins) produced. The properties of the silks are described across species, so we know that:

① The spidroins form crystalline and non-crystalline nanostructures that respectively contribute to the silk's strength and extensibility;

② The amino acid composition of the spidroins correlates well with certain nanostructures. Dragline silk is manufactured in the major ampullate gland, which consists of three subsections that serve spidroin production, storage of the liquid precursor (dope) and fiber formation respectively [Fig.3.2(a)]. Prior to extrusion, the dope flows through a funnel-shaped aperture, and the decreased lumen width generates shear stress on the dope, inducing a thinning and solidification of the fiber. Despite a good working knowledge of silk genetic structures and an understanding of the influences of genetic expression on the proteins produced and the functional properties of the proteins, commercial-scale engineering of a material that performs as well as natural spider silk has proved elusive.

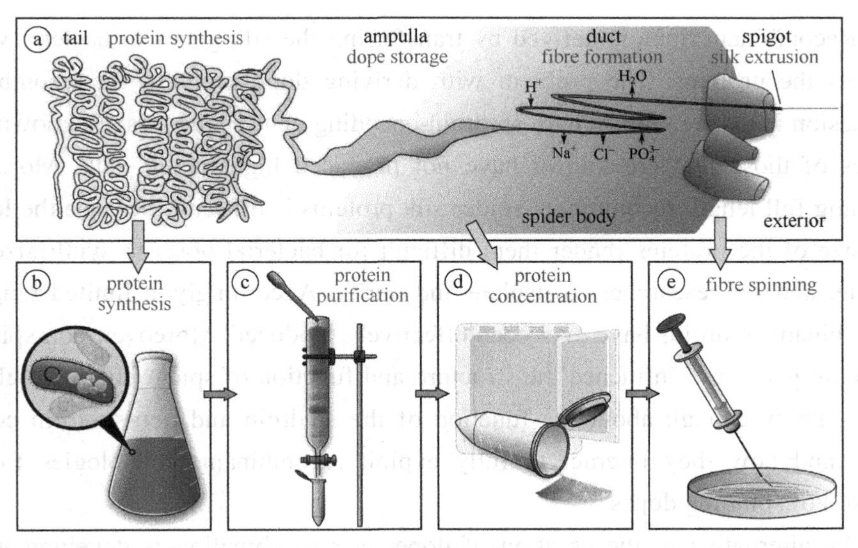

Fig.3.2 Production of spider silk. The dragline silk of orb web spiders is produced in the major ampullate gland, located in the abdomen. The gland is built like a production line (a), with the tail cells synthesizing the silk proteins (spidroins), the ampulla (or sac) storing large quantities of the spidroins in a solution called dope, the duct forming a fiber from the dope by shear forces and ion exchange and the spigot extruding the silk to the exterior. To synthesize large quantities of spider silk for industrial and biomedical applications, a biotechnological approach is used, where these different tasks are performed separately and additional steps are necessary [(b)-(e)]. To produce the base material—the spidroins, parts of the silk gene are transferred into bacteria, which express the so-called recombinant spidroins in their cells (b). The cell suspension is then dissolved and the recombinant spidroins are separated by column chromatography (c). The recombinant spidroins are concentrated (d), and may be subject to further chemical treatments. There are different methods to spin the recombinant spidroins into fibers, of which the most common is to extrude them through a tapered syringe into a buffered saline solution (e)

3.2.2.2 Biomimetic Approaches to Spinning Synthetic Spider Silk

The development of synthetic spider silk involves the creation of spinning dope and a biomimetic spinning process [Fig.3.2(b)-(e)]. The biomimetic spinning process includes chemical and physical treatments of the proteins under specific conditions to promote aggregation and folding of the proteins at precise moments as well as controlled drawing of the solid fiber. There are limitations to the effectiveness at each stage.

3.2.2.3 Creating the Proteins and Their Treatment

Three alternative sources of spinning dope have been utilized: native, reco-mbinant and genetically modified. Native dope contains the desired proteins in the desired ratio, so is considered the “gold standard” used to test the efficiency of the treatments and spinning methods. Native dope is obtained directly from the glands of sacrificed spiders or from spun fibers dissolved in caustic solvents.

Recombinant dope is derived by transferring the silk genes to bacteria which express the proteins. The problem with deriving dope proteins via recombinant expression is that few full-length spidroin-encoding gene sequences are known, and clones of those that are known have not produced high quality silk. Moreover, attaining full length recombinant spider silk proteins is difficult, because the length and size of the proteins render them difficult for bacterial hosts to synthesize and secrete, and for researchers to isolate and purify. Accordingly, a limited range of recombinant spidroins have ever been effectively produced. Moreover, non-spidroin encoding genes can influence the structure and function of spider silks. We clearly do not know enough about the function of the spidroin and non-spidroin coding genes and how they interact to fully exploit recombinant technologies for the creation of spinning dopes.

An alternative to the creation of dope by recombination is development by genetic modification of the proteins secreted by bacteria (e.g. *Escherichia coli*), plants (e.g. tobacco), yeasts, or animals (e.g. goats). This approach produces dopes that contain silk-like proteins but of significantly reduced molecular weight. These proteins accordingly require further chemical treatment to produce silk threads.

3.2.2.4 Biomimetic Spinning

A combination of changes in water content, salt concentration, pH and shear stresses act on the dope as it flows from the sac to the duct during natural silk spinning, and these induce the proteins to rearrange into nanostructures that give dragline silk its properties. Accordingly, immersion in a combination of saline and acidic solutions is utilized prior to and during biomimetic spinning. An inability to precisely co-ordinate the actions of salts, pH and shear stresses have generally led to the synthesis of inferior artificial silks.

The methods used for spinning artificial silk fibers currently include the pulling of fibers through microfluidic, electrospinning and mechano-spinning devices at draw speeds as close as possible to natural spinning speeds. During spinning, the crystalline and non-crystalline nanostructures self-align to varying degrees depending on draw speed and frictional forces at the spinning valve. The faster the draw is, the greater the nanostructural alignment and the greater the stiffness of the silk is. Researchers have, however, so far only spun silk fibers with strengths and extensibilities, about half the value of native dragline silk using recombinant proteins spun into a water solvent.

3.2.2.5 Evaluation

As indicated, each step in the production process of synthetic silk fibers exhibits problems that are still challenging despite a relatively good understanding of the natural silk secretion process. It is apparent that we cannot simply copy the elaborate synthesis and secretion process of a gland into a biotechnological process. Recombinant protein expression and artificial spinning are an inevitable necessity to produce fibers with properties that mimic those of spider silk. When developing a synthetic spider silk, we should also ask: what are the properties we desire and why? If, for instance, the desired properties do not match those of dragline silk, then some refinements of the current spinning methods might suffice. Ultimately not all properties of dragline silk will be desirable in commercial materials. For instance, dragline silk shrinks and becomes rubbery when exposed to water. While this property might be useful in some contexts, clearly a rigid structural material should not have this property and the proteins or spinning processes might need to be further modified to remove it.

3.2.3 Worms Show Way to Efficiently Move Soil

By passing small electric currents between electrodes on a surface, it is possible to attract water lubrication to points which most need it, and reduce resistance to passage for bulldozing blades and agricultural ploughs by a very significant amount. The discovery has come from a Chinese study of how it is that earthworms and dung beetles ease their path through the soil. The amounts of electricity required are microscopic—earthworms only need to apply a few tens of millivolts, increased to 12 V for machines working on a man-made scale. The discovery is the work of a team of researchers led by Professor Luquan Ren at Jilin University.

3.2.3.1 Bioelectricity in Living Creatures

If electricity is passed through soil, water tends to migrate from positive to negative poles under the action of a surface double layer effect called electro-phoresis or electro-osmosis. Bioelectricity is found in all living creatures. It is of two types. There is a resting potential between the inside and outside of the tissue or cell membrane when the creature is stationary. When moving, there is an additional action potential between the excited part and the resting part of the same tissue or cells. The action potential is of short duration, but larger than the resting membrane potential. Surface potentials on the outer skin of an earthworm arise from the resting and action potentials.

In the study in China, surface potentials on the fore part, middle part and hind parts of an earthworm were measured using silver-silver chloride electrodes and silk electrodes. It was found that the resting potential of the earthworm was zero with respect to earth, but the skin of the moving part rose to as much as −40 mV at the fore part of the worm with respect to earth and to the rest of the worm. In another experiment, a circular silver-silver chloride electrode was prepared and fixed inside a 10 mm long tube. The earthworm was guided through it and it was found that the worm produced a potential of −35 mV at its fore part, −19 mV in its middle part and −18 mV at its hind part.

In addition to the generation of voltage between the front and rear parts of the worm, it has been found that in various animals which spend much of their time in contact with soil, their skins are ribbed or rough in some way. In each case, surface elements which stick out show a small negative voltage relative to areas in between. The purpose appears to be to cause water to move towards protuberances in order to lubricate them. Although the voltage differences are very small, the distances are also very small, so that voltage gradients are significant. Water is an extremely good lubricant, as former school students should remember from having to wet glass tubes in order to slide on rubber connecting tubes. Electrophoresis is also known to be the most efficient way of moving fluids on the small scale, and is generally expected to be key to pumping within the emerging generation of lab on a chip devices.

3.2.3.2 Rough Surface of Soil Animal

Dung beetles are found to use a similar mechanism to earthworms to help them make their way through soil. It has been concluded that for maximum effectiveness, it is not only necessary for voltage to be generated between different parts of the skin surface, but also that this should be rough. If it is rough, water is encouraged to migrate to relatively small areas where lubrication is of maximum benefit.

3.2.3.3 Evaluation

The Chinese team has been quick to see whether the effects would be useful in larger scale products. Starting with laboratory scale bulldozing experiments, it was found that under otherwise identical conditions, soil stuck onto the surface of smooth or rough plates without electro-osmosis, but little or no soil stuck to a rough plate with energized electrodes. Dome topped electrodes in the test plate were made from carbon steel, placed in holes in the test plate within insulating containments. Dome tops, which stood slightly proud of the flat plate surface were energized at −12 V relative to the plate. The team reports that the arrangement reduces bulldozing resistance by 15%-32%. The team then went onto tests with modified plough mouldboards in the field. It was found that tillage resistance was reduced by 15%-18%, and fuel consumption reduced by 5.6%-12.6% under the same conditions of field, tractor, operator and weather. After 500 hours' use, the dome tops showed some signs of abrasion while the edge of the mouldboard showed "notable" attrition. Hence it would appear that improving lubrication of the surface by inducing additional water based lubrication also reduces wear. Trials have been undertaken to investigate possible improvements that the method might achieve in the functioning of coal hoppers in electricity generating stations, and steel chain liners in dump trucks, and flexible steel liners in loader buckets.

3.3 Pathway to Enhanced Outcomes

3.3.1 Specification of the Target Function(s)

(1) Defining the objective

The starting point of any biomimetic approach is a key question. This may be an unresolved technical problem (top-down approach) or the aim to understand a biological function (bottom-up approach). In both cases, the problem should be clearly defined. For example, it might be asked how we can achieve enhanced toughness in a certain material to prevent fracture under defined load, or how does an aquatic insect retain the air bubble around its abdomen when underwater. However, in many cases the problem may be more complex. For instance, when asking how can we enhance the toughness of 3D-printed components or how do geckos reversibly stick to various surfaces.

(2) Defining elemental functions

If the aim is multi-functionality as in gecko-inspired adhesives and synthetic spider silks, then a more directed approach is to break down the problem into single, clearly defined sub-topics. Sub-topics represent single functions that are, at first, studied separately, and later jointly implemented into the product. Such a modularization of the core problem may facilitate the identification of elemental working principles and guide the subsequent design process. For instance, in spider silk it was found that extensibility and strength are caused by specific motifs within the amino acid sequences. Such motifs can be used as "building blocks" to design tough silks that are much simpler than the natural model and can be tailored for the intended application.

(3) Defining the intended application

In the above examples, the final applications are only vaguely defined, possibly because much of the work is purely exploratory. In the case of gecko toe biomimetics, a universal reversible adhesive will be difficult to achieve and will always involve a compromise between conflicting. Instead, clearly defining the specified strength of adhesion required and to what specified substrate or surface it is to be applied will assist the development of more directed biomimetic procedures.

3.3.2 Choice of Model

If the target function is clearly defined, the outcomes may depend on the suitability of the model. This requires a basic understanding of the biological role of the target function within the model system. For instance, the adhesive system of the gecko toe is adapted to rapid movements, rather than strong attachment, and most applications may require durable, strong adhesion. Accordingly, the design features of one of the few commercially successful structural adhesive tapes were found in the adhesive system of male leaf beetles that durably attach to the smooth wings of their female partners. Since engineers, and even biologists, will not know the vast biodiversity and biological literature, a beneficial tool would be a central searchable database that gathers biological model systems and working principles in a uniform style.

3.3.3 Extraction of Working Principles

Understanding the working principles and their isolation from the biological model (in some literature, this process is called "abstraction") is a crucial and risky step because it is unpredictable. As we illustrated at the outset, multi-functional trade-offs and evolutionary history can obscure structure-function relationships and make it difficult to identify the basic unit that is responsible for the function of interest. Furthermore, the function of interest may be caused by a set of features, some of which might not be apparent. For example, it has been largely overlooked that the properties of the underlying soft tissue have a major effect on the functions performed by the surface features of gecko toes and shark skin, which has led to some false assumptions.

For the identification and extraction of working principles, the following approaches have proved useful and led to success in exemplary cases.

(1) Experimental manipulation

Descriptive performance assessments are typically used to identify copy-worthy features. However, such assessments will not uncover working principles. Where possible, the most efficient way to identify working principles is to experimentally deactivate features that are putatively involved in the target function and to observe the effects on the target function. However, it is not always possible to disable single features. For instance, in geckos, setae could be removed or sealed, but it is not possible to manipulate the stiffness gradient or chemical composition of the setae.

(2) Computational models

Computer models provide an alternative method for identifying the functional relevance of traits when the experimental testing of functionality is difficult. For instance, simulations using finite elements and similar models have been used to test the effect of different features of adhesive setae on their adhesive properties, like stiffness gradients, or contact geometry.

(3) Use of comparative methods

One of the most reliable ways to identify the functionally relevant aspects of a trait is to reconstruct the trait's evolutionary history using comparative methods. For example, where a trait evolved once in a common ancestor of a species group and was lost by members of the group which do not perform a function of interest but retained by other members of the group which do, we have evidence that the trait evolved, at least in part, to perform the relevant function. Even more informative are cases where a similar trait and function has evolved multiple times independently among species, thus helping to decouple the effect of a function from the effect of ancestry.

(4) Artificial models

Simplified, designed models are useful tools to test hypothetical working principles, because they permit considerable freedom in altering various parameters. This has repeatedly been applied in studies on gecko-inspired adhesives. In functional materials, recent advances in 3D-printing technology have opened new doors to test hypotheses on working principles with physical models from nano- to macro-scales. However, this approach risks a costly and time-consuming cycle of trial-and-error. Therefore, this method is ideally deployed once there is a basic understanding of the factors involved. Some researchers demonstrated a time-effective workflow for structural biomimetic materials, where the results of basic morphometric and mechanical measurements of the biological model are fed into a computer model that is used to find the optimal parameters. These are used as a blueprint for 3D-printed physical models to test the proposed working principle.

3.3.4 Designing Prototypes

The process of prototype design should consider the final application and scale of use. The main limitation in this step is the availability of fabrication methods. Biomimetics is especially constrained by the fact that biological systems are built additively at a nano-scale, and analogous technologies are still premature. This is one of the main obstacles for the successful implementation of synthetic spider silk production and gecko-like adhesives.

3.3.5 Testing Prototypes

The testing of prototypes against the previously defined target function and the comparison of their performance with the biological model and existing products is, regrettably, often neglected. This may be because scientists are under pressure to produce success stories. As discussed in the example of gecko-inspired adhesives, tests need to be standardized to objectively evaluate the performance of the prototype. If the prototype meets the objective, next steps will involve its implementation into a commercial product.

Key vocabularies

abdomen [ˈæbdəmən] **n.** （人体）腹，腹部；（昆虫）腹部
adhesive [ədˈhi:sɪv] **n.** 黏合剂，黏着剂. **adj.** 可黏着的，黏性的
amino [əˈmi:nəʊ]**adj.** 氨基的. **n.** 氨基
ampullate gland **n.** 壶状腺；壶腹腺
ancestor [ˈænsestə(r)] **n.** 祖先，祖宗；原种；原型，雏形
ancestry [ˈænsestri] **n.** 祖先；血统；世系；演化；起源
anisotropy [əˈnɪsətrəpɪ] **n.** 各向异性，异向性；非均质性
aperture [ˈæpətʃə(r)] **n.** 孔，洞；光圈；孔径；隙缝
aqueous [ˈeɪkwiəs] **adj.** 水的，水般的，水成的
attrition [əˈtrɪʃn] **n.** 消耗；消磨；摩擦
broach [brəʊtʃ] **vt.** 提出，谈起；钻孔；钻孔取液体. **n.** 钻头；凿子
bulldoze [ˈbʊldəʊz] **vt.** （用推土机）推倒；铲平；（使）强行通过
cannibalize [ˈkænɪbəlaɪz] **v.** 吃同类的肉，同类相食
caustic [ˈkɔ:stɪk] **adj.** 腐蚀性的；苛性的
chromatography [ˌkrəʊməˈtɒgrəfi] **n.** 套色版；色层分析；色谱分析法
circumvent [ˌsɜ:kəmˈvent] **v.** 围绕，包围；规避，回避；改道，绕过；克服

compelling [kəmˈpelɪŋ] ***adj.*** 令人信服的，有说服力的，不可抗拒的
compliancy [kəmˈplaɪənsɪ] ***n.*** 顺从；依从（等于 compliance）；合规性，遵从性
comply [kəmˈplaɪ] ***v.*** 遵从，服从
corollary [kəˈrɒləri] ***n.*** 推论，必然的结果. ***adj.*** 关联的
crystalline [ˈkrɪstəlaɪn] ***adj.*** 水晶的；似水晶的；结晶质的；清澈的
deactivate [ˌdiːˈæktɪveɪt] ***vt.*** 解除动员；使无效；复员；使不活动
discoidal [dɪsˈkɔɪdl] ***adj.*** 盘状的；平圆形的. ***n.*** 圆盘
dope [dəʊp] ***n.*** 润滑脂
efficacy [ˈefɪkəsi] ***n.*** 功效；效力；效验；生产率
elaborate [ɪˈlæb(ə)rət] ***adj.*** 复杂的；详尽的；精心制作的
electro-osmosis [eˈlektrəʊ ɒzmˈəʊsɪs] ***n.*** 电渗
electrophoresis [ɪˌlektrəʊfəˈriːsɪs] ***n.*** 电泳
elusive [iˈluːsɪv] ***adj.*** 难以捉摸的；难以理解的；难以发现的，难以捕获的
exemplary [ɪgˈzempləri] ***adj.*** 典范的，优异的；严厉的；值得效仿的
extensibility [ɪksˌtensəˈbɪlɪtɪ] ***n.*** 可延长性，展开性；可延展性；延伸性
feat [fiːt] ***n.*** 功绩，壮举；武艺，技艺. ***adj.*** 合适的，灵巧的
fibrillary [ˈfaɪbrɪlərɪ] ***adj.*** 纤丝的
hamper [ˈhæmpə(r)] ***vt.*** 妨碍，束缚，限制；使困累. ***n.*** 阻碍物，足械；船具
hierarchical [ˌhaɪəˈrɑːkɪkl] ***adj.*** 按等级划分的，等级（制度）的；分层的
immersion [ɪˈmɜːʃn] ***n.*** 浸没，浸泡
impede [ɪmˈpiːd] ***vt.*** 阻碍；妨碍；阻止
keratin [ˈkerətɪn] ***n.*** 角蛋白；角素，角质
keratinous [kəˈrætɪnəs] ***adj.*** 角朊的，角质的，角（质）蛋白的
lumen [ˈluːmen] ***n.*** （管状器官内的）内腔
microscopic [ˌmaɪkrəˈskɒpɪk] ***adj.*** 微观的；极小的，微小的
modularization [mɒdjʊləraɪˈzeɪʃən] ***n.*** 模块化
morphometric [ˌmɔːfəˈmetrɪk] ***adj.*** 形态测定的
mortar [ˈmɔːtə(r)] ***n.*** 砂浆，灰浆
motif [məʊˈtiːf] ***n.*** 基序，模体；主题；图形
orb [ɔːb] ***n.*** 球；天体；圆形物. ***vt.*** 成球形；弄圆；围着. ***vi.*** 沿轨道运行
outset [ˈaʊtset] ***n.*** 开始，开端
pharmacy [ˈfɑːməsi] ***n.*** 药房；配药学，药学；制药业
pitfall [ˈpɪtfɔːl] ***n.*** 陷阱；圈套；诱惑；隐患
pliable [ˈplaɪəbl] ***adj.*** 易弯的；柔韧的；柔顺的；能适应的；圆滑的
polar [ˈpəʊlə(r)] ***adj.*** 极地的，磁极的；极的. ***n.*** （几何）极线
protuberance [prəˈtjuːbərəns] ***n.*** 突起；瘤；结节
putatively [ˈpjuːtətɪvlɪ] ***adv.*** 推定地
redundancy [rɪˈdʌndənsi] ***n.*** 多余，累赘；复置装置，冗余

render [ˈrendə(r)] ***v.*** 使成为，使处于某种状态；（以某种方式）表达，表现
robustness [rəʊˈbʌstnəs] ***n.*** 鲁棒性；稳健性；健壮性；坚固性
rubbery [ˈrʌbəri] ***adj.*** 橡胶似的；有弹力的；坚韧的
sac [sæk] ***n.*** （动植物组织中的）囊，液囊
saline [ˈseɪlaɪn] ***adj.*** 盐的；含盐的；咸的. ***n.*** 盐溶液；生理盐水
scanning electron microscopy ***n.*** 扫描电子显微镜检查法；扫描电镜
secrete [sɪˈkriːt] ***vt.*** [生]分泌；隐匿，隐藏
seta [ˈsiːtə] ***n.*** 刚毛，刺毛（复数: setae ['siːtiː]）
silver-silver chloride ***n.*** 银-氯化银电极
spatula [ˈspætʃələ] ***n.*** 匙形，匙状；铲，漆工抹刀
spatulae ***n.*** 铲状匙突
spatulate [ˈspætjʊlɪt] ***adj.*** 竹片状的. ***vt.*** 用刮刀（或药刀）涂敷（或调拌）
spidroin ***n.*** 蜘蛛丝蛋白；蛛丝马迹
spigot [ˈspɪgət] ***n.*** 龙头；栓；饮水的地方
suffice [səˈfaɪs] ***v.*** 足够，足以；满足……的需求；有能力
tensile [ˈtensaɪl] ***adj.*** 可拉长的，可伸展的；张力的，抗张的
tensile strength ***n.*** 抗拉强度，抗张强度，拉伸强度；拉伸力
tilting [ˈtɪltɪŋ] ***n.*** 倾卸台. ***adj.*** 倾斜；倾卸
tissue [ˈtɪʃuː] ***n.*** （动物或植物的细胞）组织
trade-off ***n.*** 平衡，协调；妥协，让步；交易；权衡
trait [treɪt] ***n.*** （人的个性的）特征，特点；遗传特征；一点，少许
trial-and-error ***n.*** 反复试验法，试错法；不断增加；不断摸索
unviable [ʌnˈvaɪəb(ə)l] ***adj.*** 不能独立生存的；不可行的
van-der-Waals force ***n.*** 范德华力，范德华引力
wedge [wedʒ] ***n.*** 楔子，三角木；楔形物

Exercises

Ⅰ. Translate the following English into Chinese.

1. In effect, biomimetics assumes that evolution by natural selection is a series of natural experiments that have optimized a design and rendered any suboptimal alternatives extinct. Biologists question this assumption, highlighting the difficulties that may arise because of systematic differences between evolution by natural selection and engineering procedures.

2. In order to design gecko-inspired adhesives from synthetic polymers, it has been essential that engineers understand the basic physical mechanisms acting in gecko toe pad surfaces. Researchers have used techniques such as scanning

electron microscopy to examine the microscopic hair-like keratinous protuberances called setae on the gecko toe.

3. The methods used for spinning artificial silk fibers currently include the pulling of fibers through microfluidic, electrospinning and mechano-spinning devices at draw speeds as close as possible to natural spinning speeds. During spinning, the crystalline and non-crystalline nanostructures self-align to varying degrees depending on draw speed and frictional forces at the spinning valve.

4. Where possible, the most efficient way to identify working principles is to experimentally deactivate features that are putatively involved in the target function and to observe the effects on the target function. However, it is not always possible to disable single features.

5. Simplified, designed models are useful tools to test hypothetical working principles, because they permit considerable freedom in altering various parameters. This has repeatedly been applied in studies on gecko-inspired adhesives. In functional materials, recent advances in 3D-printing technology have opened new doors to test hypotheses on working principles with physical models from nano- to macro-scales.

Ⅱ. Answer the following questions based on the main text in this chapter.

1. What are the systematic differences between evolution by natural selection and engineering procedures?

2. Why is there immerse interest in developing adhesives that mimic the properties of gecko toes?

3. Please list the problems of developing adhesives inspired by gecko toe pads.

4. What are the three alternative sources of spinning dope for the development of synthetic spider silk? And what are their major components, respectively?

5. Please list the approaches proved to be useful and led to success in exemplary cases for the identification and extraction of working principles.

Ⅲ. Writing.

Assuming you are a salesman of the company Sto SE &Co, you are required to write a reply letter to a customer, named Tom, about his inquiry of the product of Lotusan® facade paint, including the introduction on its general function, specifications, history development of this product.

Chapter 4

Biomimetics of Motion

Designers and engineers constantly strive for optimization, improving man-made artefacts in their functionality and shape. Whether in the initial or the final stages of creation, designing buildings or micromechanical devices, the architects of things choose among combinations of endless possible solutions. This process of selection is intuitively carried out, relying on experience and a "mental archive" of observed combinations. When inspiration fails, we learn to fall back on pretested solutions: this is where the advocates of biomimetics turn to nature. The living systems we have before us are in fact fine-tuned combinations of design features that evolution has tested and adapted over millions of years of patient design.

This study focuses on the features of animals and plants that enable motion, relying on the scientific work of zoologists and botanists. The aim is to identify patterns in the combinations of kinetic features, which could be a great asset for any engineer, architect, designer, or artist who wishes to integrate motion in their work.

4.1 Biomimics for Adaptivity

As evolution has shown that long-term survival depends on adaptive capacity rather than specialization, which is instead used to develop transient features aimed at temporary adjustment (on an evolutionary timescale) to specific conditions. How can we design for better adaptivity?

4.1.1 Micro and Macro

Our perception of what is perceived as micro or macro depends upon a reference in relation to which we define it. Changing the reference, its context or scale leads towards a modified notion regarding what these terms and their link are. Something that was micro in one context may become macro when related to a different scale reference.

From an engineering point of view, adaptivity in artefacts can be achieved on two different scales: on a micro- and on a macro-scale.

In the first case, micro-scale flexibility acts through the engineering of materials and of their behavior on a molecular and nano scale. This field of expertise mainly pertains to chemistry, material engineering, biology and genetics.

In the second case, macro-scale flexibility is achieved through the design of a product and its parts. A great number of fields of expertise are touched by this topic, in particular all fields related to engineering, product design, architecture, mechanics, etc.

Adaptivities on a micro- and macro-scale are however interrelated and complementary. In nature's constructions, form and function are integrated in a same system, avoiding maximization of individual functions (which might harm other functions), rather optimizing the whole. There is an extremely refined combination of high-tech and low-tech solutions in laws of nature, complementing each other. Technology also resembles nature in this aspect: a building for instance is a sum of low- and high-tech solutions working together and resulting in better quality and efficiency than if only one type of component was used.

This work aims to address adaptivity on a macro-scale. In order to do so, however, inspiration and research are conducted across scales as from a geometric and functional point of view, the differences between scales blur and become synergic.

Adaptivity on a macro-scale, through design, can be achieved mainly in two ways:

— through disassembly;

— through motion.

The difference in adaptation through disassembly and through motion is primarily the time-factor. While flexibility through motion is a response to immediate pressures, flexibility through disassembly allows to prolong the use of a system or of its parts, and to adapt over a longer timeframe.

4.1.2 Adaptivity through Disassembly

Although design for disassembly (DFD) was initially driven by economic aspects linked to the optimization of the production process, the field is today realizing the huge potential of these techniques to transform products' end of life and optimize reuse, remanufacturing and recycling of materials, components and sub-assemblies. In a world with finite resources, growing population and consumption, this topic is becoming of critical importance as all our products are quickly being outpaced and exchanged for newer, better performing devices.

A product can be disassembled either by destroying the product and recover useful parts or materials, or by reversing the assembly process. This second type of disassembly can be seen as an adaptive capacity through a process, rather than an adaptation of the organism itself.

Reversing the assembly operations can many times be a difficult task and requires most products to be designed for it. The benefits of efficient disassembly are however important allowing to consistently reduce waste, provide ready components for future re-assembly of other products, simplify recycling of unused parts, enhance maintainability, repair and substitution of obsolete parts, and thus to potentially prolong the usage of the product.

4.1.3 Adaptivity through Motion

Kinetics is in the first place a way to introduce function and adaptability in our objects to keep up with the changing human patterns of interaction with the built environment. Most man-made artefacts are designed to move. If we look at the objects surrounding us, a great majority of our tools embed motion in their design in at least one of their variants: purses have openable zips and flaps, pens are twisted or pushed to uncover the ballpoint, paperclips bend, windows rotate on hinges as books do on their spine, etc.

Moving components accommodate specific ranges of activities by changing geometry or orientation. On one hand, this gives the objects their function and usability; on the other, it makes them subject to use and wear. When motion is made impossible, the usability of the object declines and the article becomes waste, hence the need of integrating adaptivity through disassembly together with adaptivity through motion. These two concepts can together achieve change, agility and liveness together with a broad range of benefits as aesthetics, function, flexibility, energy use, reliability, user friendliness and greener life cycles in our artefacts.

4.2 Motion versus Change

4.2.1 Motion as Change on a Specific Time-scale

Movement or change in Nature can be found in three different time-related forms:

① Slow change—As is evolution, acts on a very long timescale. In a biomimetic perspective, this kind of change is rather analyzed to allow a deeper comprehension and optimization of the system's design.

② Moderate change—As for growth and decay, which embraces the lifespan of an organism. In the biomimetic perspective, these aspects are of interest when taking into account the grey energy or the life cycle of a given man-made product.

The two above-mentioned timescales of change are not appropriate models in the context of designing motion, for three main reasons:

- The timescale required for obtaining any change is too vast to allow any perception and active reaction to the change. For a change to be rapid enough to classify as motion, it must be perceived.
- Change of form through evolution, and hence also through growth and decay, implies creation (through multiplication or addition) and destruction of material, implying obvious technological limits.
- Change is not reversible and hence not repeatable, while motion is based on the repeatability of the physical change.

In order to address motion, biomimics must therefore turn to a specific timeframe:

③ Fast change—As through muscular, hydraulic, pneumatic action. This change is instantaneous, and achieved by internal or external manipulation. This category of change is in the center of the following chapters and sets the basis for a direct biomimetic study in view of a translation of some systems into a technological context.

4.2.2 Tropic and Nastic Movements

Plants are of extreme interest in the context of biomimetic kinetic design, as the mechanisms are generally less complex than other organisms in their geometry and mechanics, although many of the reactions are still carried out through complex chemical reactions.

Tropism indicates a slow adaptive motion of plants, generally achieved through growth, in response to different environmental triggers where the direction of the motion is strongly dependent on the direction of the stimulus.

- Phototropism reacts to light, as plants grow in the direction of the light source;
- Geotropism reacts to gravity, where roots and stem grow in opposite directions, when possible the first downwards and the second upwards;

More rapid tropic movements use activating and control mechanisms as pneumatic and hydraulic pressure to achieve higher flexibility:

- Heliotropism reacts to the motion of the sun, as in sunflowers, which move by twisting the stem;
- Thermotropism reacts to temperature, as plants grow towards or away from heat sources;
- Hydrotropism reacts to the presence of water, as when plants' roots grow seeking water;
- Haptotropism or Thigmotropism react to touch, as when *Passiflora* tendrils search for support to grip and climb.

Nastic movements are faster and differ from tropic movements as they concern non-directional movement in response to a stimulus.

- Photonastic movement reacts to light (or darkness), as the leaves of plants fold when light levels are low;
- Hydronastic movement reacts to humidity, as for grass leaves which fold and curl up in wet weather conditions;
- Thermonastic movement reacts to temperature, an example is plants' leaves that can change shape depending on the cold or hot weather;
- Thigmonastic, haptonastic or seismonastic movement reacts to touch. The *Mimosa pudica* is famous for closing its leaves when touched, as also the Venus Flytrap closes on its prey when adequately stimulated.

Out of these categories, mostly nastic movements are of interest in the present context and will be further explored.

4.2.3 Locomotion

Whereas most plants achieve motion while being anchored to a fix support, locomotion used by animals implies the capacity to change location through active or passive motion. More complexity and reactivity of animal structures add many more parameters to their movements, together with a multitude of possible combinations, of which some are transversally recurring.

The medium of locomotion strongly defines the physical characters of the animal. All animals are perfectly adapted to the medium they move in, and many have evolved the capacity to do so in more than one medium: frogs for instance, have evolved limbs for terrestrial locomotion and webbed feet for aquatic locomotion. Most of these multifunctional organisms do not however have full motive efficiency in both media, where more functions need to coexist and full optimization does not appear to be possible. Also, many animals moving in two media have separate sets of tools (limbs or appendages in most cases) for each media: flying fishes possess both wings, feet and tail while birds have both wings and legs.

In the matter of terrestrial locomotion, most animals have evolved motion based on the displacement of the opposites of the body, either limbless (as serpentine locomotion or concertina) or with limbs (based on the motion of an even number of limbs).

In locomotion in fluids, water and air, most organisms rely on the action of an airfoil, or a broad surface as a fin or a wing, which is rhythmically pressed towards the media to transform muscular force into a vectorial movement. In fluids, a fixed surface simply oscillating back and forth does not provide any resultant movement as the resistance of the medium is equal in both directions. Therefore, the surface must adapt its geometry to different configurations to make sure to obtain thrust forward during the beat movement and no resistance during the recovering movement. The ability of molding the surface also requires the capacity to perform many different maneuvers as ascension or descension, acceleration or deceleration and not least different cruising and turning manoeuvers.

Movement and simple control of movement can be integrated in an organism's structure by considering the global interdependence of all structural elements. There are three main ways to control a kinetic transition process in a predictable and safe manner. In biological organisms, these possibilities are most often combined.

① Control of the joints. Joints allow movement in definite directions while providing stiffness to other directions, providing a base resistance for deformations. Joints may also provide some friction to resist unwanted deformations and thus allow movement control. Movement is introduced when applied forces act on a number of joints, which allow a change in geometry through different rotating movements.

② Geometrical arrangements. The geometrical use of the curvature which is an encumbrance in bearing axial loads, can however be of interest to facilitate a more flexible and predictable arrangement. Long bones in vertebrates, as in particular leg bones are slightly curved with a geometry known as Euler buckling, which is allowed to bend when stressed, acting as shock absorbers, eventually breaking to dissipate even more energy. The force resisted is low but the strain energy absorbed is manifold higher than if the bones were straight: nature prioritizes security before performance.

③ Boundary conditions. Predictable structural limit conditions can be used to the system's advantage, as in the case of the venus flytrap. This carnivorous plant uses the instantaneous geometric deformation of its leaves from convex to concave to achieve an extremely fast closing movement able to trap its preys.

4.3 Case Studies of Motion in Nature

In nature, there is maximal exploitation of morphological features in energetic and mechanical contexts. Motion strongly relies upon the geometrical features of the body parts. This is achieved through co-evolution of morphology, movement type and movement control.

Plant movements, and especially the mechanics of seed dispersal, are designed with simple structures and kinetics, compared to the complexity of animal locomotion. The absence of neural networks (or the presence of more basic ones), the environmentally triggered mechanisms and the subsequent chain reactions triggering motion, are three main aspects that make plants extremely interesting for a biomimetic technological transfer.

Types of movements used by animals, both in locomotion and for a wide range of movements are divided into mechanics of soft-bodied systems and mechanics of rigid systems.

4.3.1 Plants and Seeds

Plant movements are mostly based upon an exchange in fluids within the plant-swelling up upon liquid intake and shrinking when undergoing water loss—regulating the internal pressure (turgor pressure). Most tropic movements are carried out based on this type of mechanics, as the sunflower following the sun path, leaves opening and closing for day/nighttime. Plant cell's pressure can be regulated actively (controlled by the plant), or passively (controlled by the environmental conditions, as drought).

All kinds of movements in plants can be placed into two main categories.

① Slow hydraulic movements;

② Rapid movements due to the release of built-up elastic potential energy.

Rapid motion in plants is based on the release of stored elastic energy and is mainly used for seed and sporangium dissemination, defense and feeding. Among the techniques used in rapid movement, two sub-categories can be identified:

- Snap-buckling geometries;
- Explosive and fracture dominated geometries.

The advantage with the former one is the reversibility of the mechanism, which switches between two opposite and complementary shapes, not involving the tearing of the plant tissue to release the stored energy. On the other hand, the fracture-dominated mechanisms are generally also the fastest ones. Most often, these two kinds of plant movement can be found to work together, while completing each other: the hydraulic mechanisms build up energy in the structures, bringing them to the limit of structural stability, and the fracture or geometrical snap-buckling instantly releases the energy.

The case-study sheets explore the following vegetal systems, with a specific focus on one of its kinetic features (Table 4.1).

Table 4.1 The specific kinetic aspect of vegetal systems studied

Name of the organism	Kinetic aspect studied
Morning glory	Photonastic flower opening
Mimosa pudica	Thigmonastic defense
Venus flytrap	Thigmonastic hunting

4.3.1.1 Morning Glory, Photonastic Opening and Closing Hydraulics

Coordination of synchronized flower opening is often a quite complicated molecular mechanism, working in parallel with the normal growing processes of the plant. The timing of opening is precisely regulated by factors as light intensity, temperature and humidity for the nocturnal species. The triggering factor generally needs to act under certain duration before the flower opens, avoiding errors due to the interaction of other factors. Closing of the petals can be due to aging (giving permanent closure) or to an active process similar to the opening one. The *Ipomoea*, also called morning glory, is native to tropical and subtropical regions. The floral petal opening is quite fast, occurring within a range of minutes and is mainly due to a phytochrome reaction, responding to the changing light conditions making the flower opening at dawn (Fig.4.1).

Fig.4.1 Three shapes in the *Ipomoea* flower (from left): closed bud, open corolla, toroid bud closure. In the fully opened flower, the midribs are visible in a lighter color under the flower membrane

(1) Structure and geometry

The *Ipomoea* petals are helically furled and densely packed in a 10 cm long, 1-2 cm wide bud. The surface of the flower is made out of a single umbrella-like membrane wrapped 540° clockwise around the central axis in a logarithmic helix. Petal unfolding occurs within a few minutes in an outward spiraling opening, the final position of the petals depending considerably of their position in the bud. The unfolded flowers bloom into an 8-15 cm radial pentameric umbrella (the lamina) with a tube-shaped base.

The size of the corolla is strongly influenced by the amount of turns in the helix, the more turns, the wider the flower. For this reason, flowers with constant spiral folding can potentially generate larger flowers, although the logarithmical helix packaging provides more stable folded geometry (Fig.4.2).

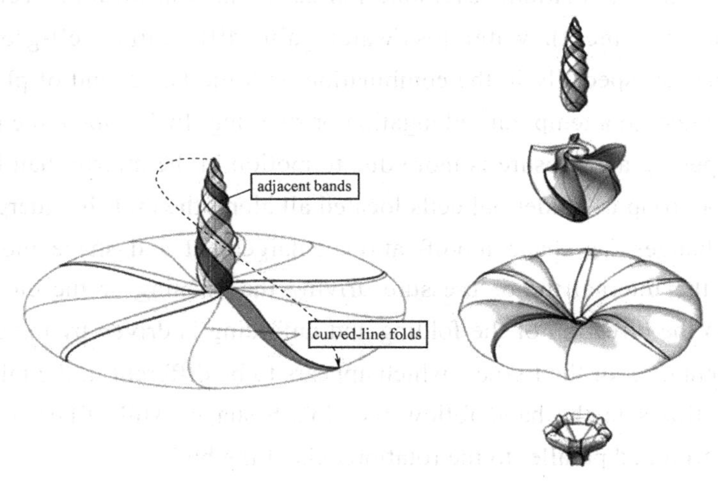

Fig.4.2 Geometry and opening sequence of the *Ipomoea*

A stiff band running along the petal's symmetrical axis—the midrib, characterizes the underside of the petals. In the bud state, the rolled up geometry exposes these bands outwards, the ridges running adjacently to each other and effectively stiffening the bud and forming a shell-like protection. In the unfolded corolla, the band stretches the surface wide open and helps supporting the spanning structure by forming a large cantilever funnel stabilizing the surface through its double curvature rolled beam able to carry the load of the pollinators landing on it. Furthermore, each rib displays an S-shaped curvature along its length, providing much more stiffness than a linear beam.

(2) Motion sequence

Two different opening motions can be identified in the morning glory flower: a helical unrolling around the central axis, and an inwards curling of the petals. In the first case, the unfolding sequence of the densely packed bud depends on a contemporary helical unrolling of the structural bands and the unwrapping of the petal surface. Not being able to use the geometrical overlapping between petals to open the flower, the *Ipomoea* undertakes several complex structural changes folding, bending and expanding the corolla.

In the second case, the flower closes at sunset by curling the surface inwards by more than one complete turn. This vertical curling is fully reversible, as the mechanical tension in the petal drops as consequence of lower light conditions, the bud undertakes a toroidal shape.

(3) Motion mechanism

Opening of flowers depend on four different kinds of petal movements which mostly occur in combination: reversible ion accumulation, localized cellular death (irreversible movement), water loss/water gain, differential cell-growth. Most characters occur especially in the combination with the latter kind of phenomenon, as the cells undergo a temporary elongation or swelling. In *Ipomoea*, the mechanism of flower opening and closure is more due to motion in the midrib than in the petal lamina. As a group of epidermal cells located all along the midrib undergo dynamic structural changes (as shape modification, enlargement and shape modification), they affect the internal turgor pressure driving the opening or the closing of the flower. Also the direction of the folding and unfolding is driven by the distribution and cell orientation of the tissues, which appears to be different in the midrib and in the lamina: fibers in the band follow the rib's S-shape, while fibers on the petal surface are arranged parallel to the rotation axis of the bud.

4.3.1.2 *Mimosa pudica*, Defensive Nastic Reaction in the Leaves

Mimosa pudica L. (Mimosaceae), also called "touch-me-not", "shame" or "humble plant", is a sensitive plant responding to mechanical stimuli with a rapid seismonastic or thigmonastic reaction of its leaflets. Evolution of this trait is thought to balance the risk of being eating by herbivores, at the expense of a substantial energetic cost and interference with photosynthesis (Fig.4.3).

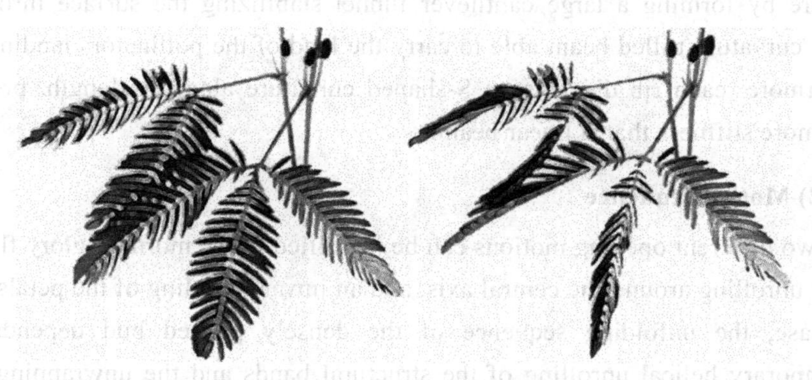

Fig.4.3 A *Mimosa pudica* before (left) and after (right) triggering the thigmonastic reaction

(1) Structure and geometry

The *M. pudica* is a prostrate or semi-erect bushy plant which grows up to 1 m with about 12-20 pairs of bi-pinnate palmated sensitive leafs, each in turn composed of 10-20 pairs of 0.6-1.2 cm long, 0.3-0.4 cm broad oblong glabrous leaflets with acute apex and rounded base, also called "petioles".

(2) Motion sequence

Movement can be initiated by many different kinds of interaction: touch, blowing, shaking, heat, rain, light, etc. Generally the movements observed are of two different kinds:

- A slow, nyctinastic, periodical movement controlled by the plant's biological clock in response to daylight and night hours;
- A very rapid reaction in response to punctual stimulation or stress.

In the second case, action-reaction is immediate: the leaf rapidly folds completing the reaction in 1～2 s. If no further stimulation occurs, the leaf opens up again after a few minutes. The more intense the stimulation have, the longer the interval is before opening (Fig.4.4).

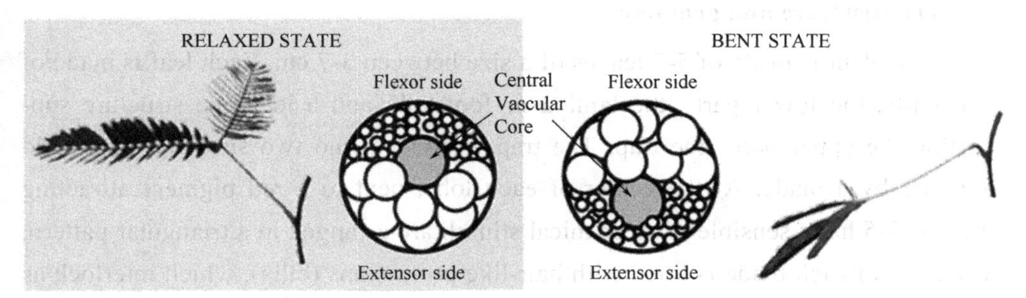

Fig.4.4 Schematic diagram of the section of a pulvinus in the relaxed state (before stimulation), and during bending (after stimulation)

(3) Motion mechanism

The force-velocity ranges of the contraction mechanism has been compared to that of the typical animal muscle movements and have been found to be not only remarkably similar, but actually more efficient in the mechano-chemical energy conversion process.

The bending movement is carried out by the pulvinus — the joint-like thickening at the base of the leaflets—and not by the petioles themselves. When disturbed, specific regions of the plant's stem release chemicals, which activate so called "motor cells" located in the lower part of the pulvinus. The initially swollen cells undergo a drastic change in volume, collapsing through a temporary loss of turgor pressure, with a consequent curvature of the pulvinus in the part where these cells are located, causing the folding of the leaf.

4.3.1.3 Venus Flytrap, Thigmonastic Hunting Snap-Trap Mechanism

The venus flytrap (*Dionaea muscipula*) is a carnivorous angiosperm, which has evolved its leaves as snap traps at the expense of part of its photosynthetic capacity. The snapping motion is one of the most rapid movements in the plant kingdom, closing in about 0.3 s at a speed of 100 m/s (Fig.4.5).

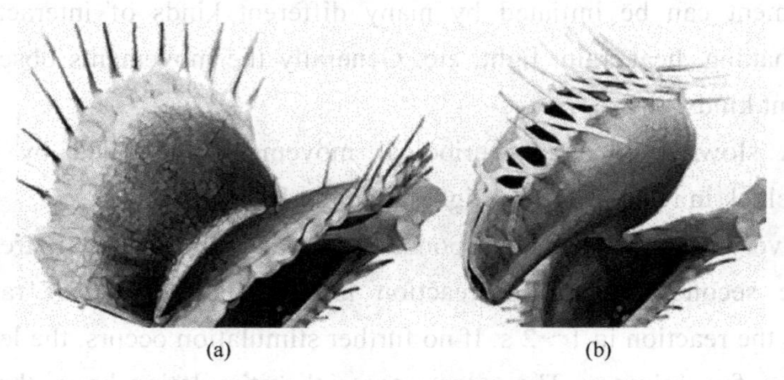

Fig.4.5 A *Dionaea muscipula* in the open trap phase (a), and in the semi-closed trap phase (b)

(1) Structure and geometry

The plant is made of 5-7 leaves of a size between 3-7 cm. Each leaf is made of two parts, the lower part—the lamina or footstalk, and a leaf like structure supporting the upper part—the trap. The trap is divided into two specular lobes held together by a blade. At the center of each lobe, next to a red pigment attracting insects, 3-5 hairs sensible to mechanical stimuli are arranged in a triangular pattern. The edge of each blade is lined with hair-like protrusions (cilia), which interlock as the leaves closes, preventing all but the very smallest prey to escape.

(2) Trapping sequence

① Open state, is a silent and stable phase with no observable movement. As an insect lands on the upper surface of the leaves and rubs against the trigger hairs, the trap enters an activation phase. The hairs need to be stimulated at least two times within 30 s to reach the threshold value for the trap to close. This prevents the plant to close when triggered by rain or wind, which are typically conditions below the stress range.

② The trap snaps shut in a fraction of a second, or in fact into a semi-closed state—lightly ajar, in a state between opened and closed, for several minutes. This "decision-making" phase is adaptively advantageous, preventing the high energetic costs of secreting digestive enzymes into an empty trap. At this point, two different outcomes are possible. If nothing or non-nutritive material has been captured and no further stimulation of the trigger hairs occurs, the trap returns to the fully open state in 1-2 days. However, if a prey has been captured and continues moving stimulating the trigger hairs, the trap proceeds to the fully closed state.

③ Closed state, with a new geometric equilibrium. The lobes seal tightly and remain closed for up to two weeks, during which the prey is digested.

④ Reopening of the trap for the next victim. During the loading of the snapping mechanism, the leaves change shape in a very slow process—from flat, to concave and finally to convex (Fig.4.6).

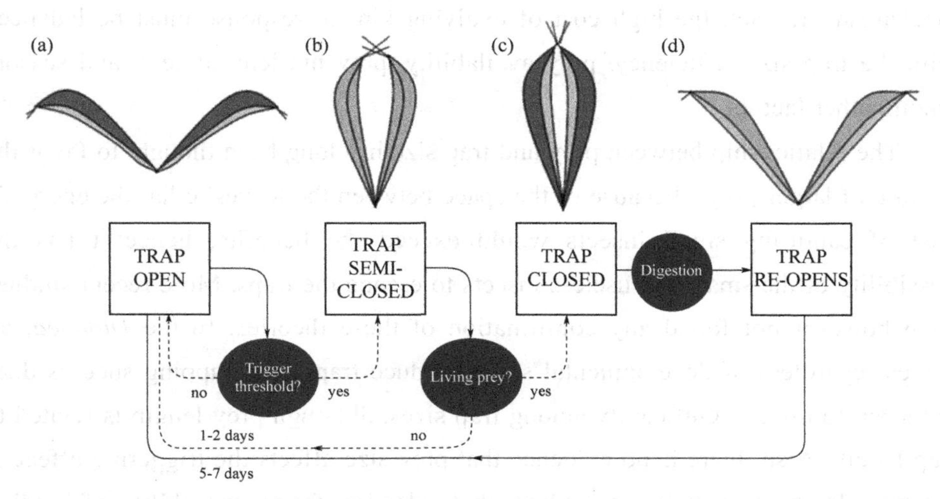

Fig.4.6 Conceptual diagram of the snapping phases and the turgid layers of the leaves. In the first phase, the inner layer is thicker than the outer layer; in the second phase, both layers have equal volume; in the third phase, the outer layer is thicker than the inner layer; in the fourth and last phase, the surfaces are minimal with zero mean curvature

(3) Trapping mechanism

Although this plant has been object of study and fascination for over two hundred years, the mechanism behind its fast nastic movement is still poorly understood. The most recent theory (the snap-buckling theory) suggests a combination of a geometric activation mechanism (a shape and curvature deformation resulting from a snap-buckling instability) and a chemical reaction (with a sudden change in cells'volume through a hydro elastic curvature mechanism).

According to the snap-buckling theory, the driving force in the snapping is the elastic curvature energy stored in the curved pressurized leaves, loaded like a spring with a pressure difference between the upper and the lower layers of the leaves in a stable state. The stimulation of the trigger hairs opens up pores in both hydraulic layers, stability drops drastically and, in order to regain equilibrium, the convex geometry of the leaf snaps closed into its opposite concave (closed) position. The movement occurs through changes in curvature rather than movement of the leaves.

These kinds of reactions based on hydraulics and mechanics represent an alternative to muscle-activated kinetics. The differential turgor pressure is actively regulated by the plants and allows fast thigmonastic movements after a longer period of building up the pressure.

(4) Scaling

Out of about 600 existing carnivorous plant species, the *D. muscipula* is among the very few together with the *Aldrovanda vesiculosa* L., which has a nastic trapping mechanism. In fact, the high cost of evolving kinetic response must be balanced with the trap size, efficiency, prey availability, prey nutrient content, and season, among other factors.

The relationship between prey and trap size has long been thought to favor the capture of larger preys, because of the space between the leaves' cilia: the energetic cost of capturing small insects would exceed the benefits, hence it has the possibility of the small and useless insects to escape the traps. More recent studies, have however not found any confirmation of these theories. In the *Dionaea*, all leaves regardless of developmental stage produce traps and trapping success does not seem to differ significantly among trap sizes, although prey length is related to trap length. Also, there is no evidence that prey size affects the triggering effect of the trap. In the end, it is more likely that selection favors the ability of holding captured preys over allowing the small ones to escape as the whole plant benefits over any insect capture across all leaves.

4.3.2 Soft-Bodied Systems

In this part, examples of hydraulic mechanics of motion and locomotion through compartment-characterized soft-bodied structures in animals are shown.

Soft-bodied structures are used both as body-structures by the organisms as a whole, built up on hydraulic-cushion systems and moving accordingly, and as organs or body parts with hydraulic characters such as feet, tongues, trunks, etc., which are used in soft-bodied and rigid structures alike. These structures range from single, multiple to juxtaposed compartment systems.

Soft-bodied locomotion is a very diffused way for moving in insects. Not being controlled by rigid parts and defined folding points (joints), these animals are able to move in any body plane, making the movements potentially very complex.

The case-study sheets explore the following soft-bodied systems, with a specific focus on one of its kinetic features (Table 4.2).

Table 4.2 The specific kinetic aspect soft-bodied systems studied

Name of the organism	Kinetic aspect studied
Earthworm	Peristaltic locomotion
Caterpillar	Crawling and inching locomotion
Asteroid tube feed	Hydrostatic locomotion

4.3.2.1 Earthworm, Peristaltic Locomotion in the *Lumbricus terrestris* L.

As most soft-bodied animals with hydrostatic skeletons, earthworms locomote by deforming their body segments.

(1) Structure and geometry

The *L. terrestris* is substantially a cylinder-formed organism enclosing a fixed volume of liquid with a musculature arranged circumferentially and longitudinally to control the diameter and the length of the cylinder by using the incompressible proprieties of the liquid to transmit motion and to provide antagonist force to re-extend the muscles.

The tissue walls enclosing the liquid are divided into about 145 segments (septa) similar to each other, lacking localized specialization. The subdivision into septa allows localized control and variable movement patterns through the regulation of the flow of the liquid in the segments, as a sluice gate system. The body changes in length by changing diameter in the septa: the body walls can extend by more than 10% through contraction of the circular muscles (Fig.4.7).

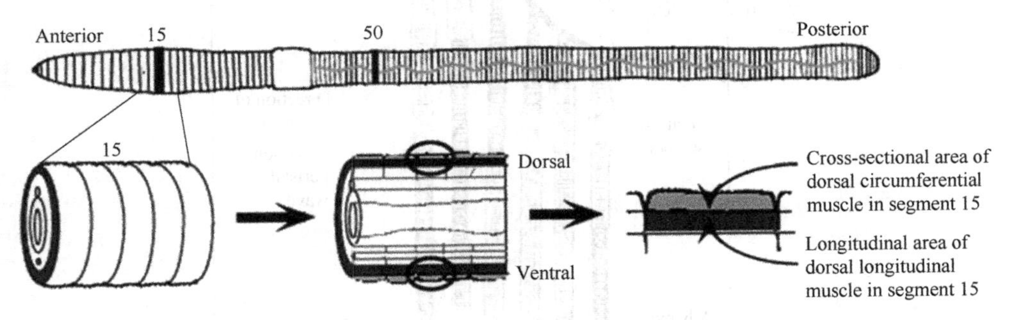

Fig.4.7 Earthworm anatomy: longitudinal section of a *L. terrestris* with zoom-in on the dorsal circumferential and longitudinal muscles

(2) Motion sequence

These soft-bodied crawlers typically use peristalsis for locomotion. Propulsion is achieved by letting rhythmic retrograde waves (waves progressing in the opposite direction to the locomotion) of muscle contraction pass through the whole body, starting from the anterior part of the animal. As a wave reaches the half of the body length, a new wave is initiated at the anterior end and the cycle is repeated over by each segment.

The expanded segments anchor the body to the substratum while the contracted ones advance, progressing with about 2-3 cm in each cycle. Speed equals the length of each stride multiplied by the frequency of the waves passing over the body.

(3) Motion mechanism

Earthworms use their internal body liquid to amplify the force, velocity and kinetic effect of their muscle-contraction by regulating the pressure in their body segments. As the circumferential muscles of a segment contract while the amount of fluid it contains remains constant, the cylinder is elongated from thick to thin. As the longitudinal muscles contract, the shape reverses to a thick and short cylinder. The pulsatile shifting between these two shapes in the septa results in a forward linear thrust of the body.

Longitudinal and circumferential muscles act as agonists: because each septum encloses a constant volume of liquid, contraction of one set of muscles results in the extension of the opposite set of muscles, force being transmitted through the fluid (Fig.4.8).

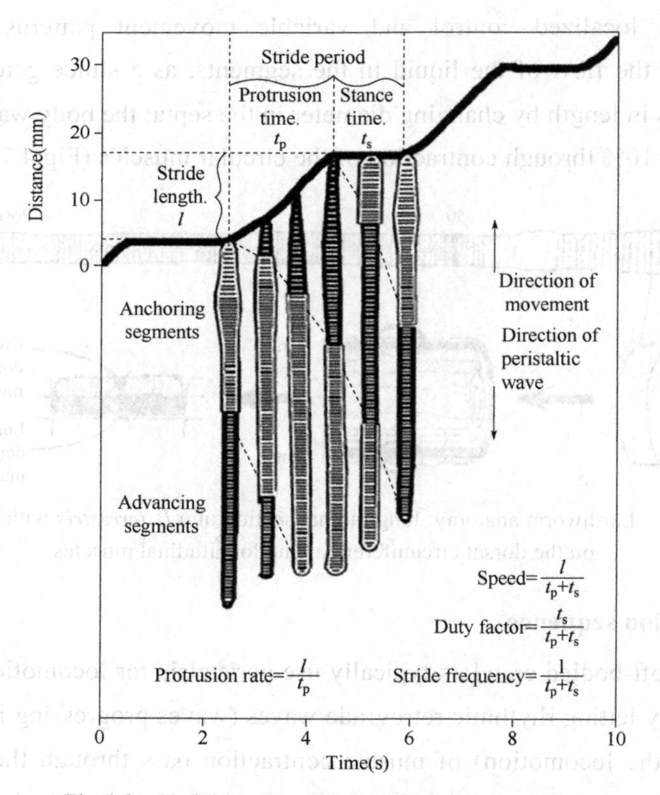

Fig.4.8　Variables describing peristaltic locomotion

(4) Locomotory characters and gait

The stride length for a hydrostat would be the horizontal distance traveled during one cycle of peristalsis. Speed can be increased by taking longer and/or more strides per unit of time. However, as speed depends on the geometrical dimensions of the body, all earthworms keep similar speeds in relation to their size. Peristaltic locomotors can change speed but do not change gait. Although speed is generally a factor of high selective pressure both for predators and for preys, in the case of crawling, this locomotory technique do not seem very fit for fast movements as organisms generally choose non-peristaltic mechanisms for fast movements.

4.3.2.2 Crawling and Inching Locomotion in the *Manduca sexta* and Other Caterpillars

Not all soft-bodied animals are hydrostats using incompressible fluids. Caterpillars have an extensive internal trachea containing, exchanging and compressing air. As the cavities are permeable and air leakages could compromise the efficiency of the force transmission, instead of losing efficiency through deformation, the caterpillars use the stiffness of the substrate to transmit forces.

(1) Structure and geometry

Caterpillars have three pairs of different legs: thoracic prolegs at the front end with a strong gripping-hook on each foot (clasper), abdominal prolegs more or less in the second half of the body-length, and anal prolegs in the rear. The body is mainly composed of one cavity without any septa or body segments. This allows body fluids and tissues to be moved along the longitudinal axis of the animal as the perimetral body segments stretch or compress, moving the caterpillar's Centre of Mass forward.

(2) Motion sequence

Caterpillar locomotion (on horizontal or vertical plane) can mainly be dived into inching and crawling. Different species of caterpillars adapt one or both techniques, supposedly in relation to body-mechanics/size: large caterpillars keep their body close to the ground while small caterpillars can achieve self-supporting stiffness. In fact, some caterpillars (not the *Manduca sexta*) have shown to switch gait during growth.

Inching involves that the worm uses only the feet on its rear end. It casts the front part of its body to find an appropriate grip, then bends in the middle of the body-length and moves the posterior part to grip next to the anterior part. The movement requires an efficient structural control, precision and coordination.

Crawling locomotion is achieved with a progressive series of anterior-posterior body-wave contractions. One cycle is characterized by three main steps:

- The tip of the abdomen is moved forward, and the terminal prolegs anchor the body to the substrate;
- An anterograde wave of muscle contraction proceeds from the rear end;
- As the wave reaches the anterior part of the animal, the legs unhook and anchor the body further away (Fig.4.9).

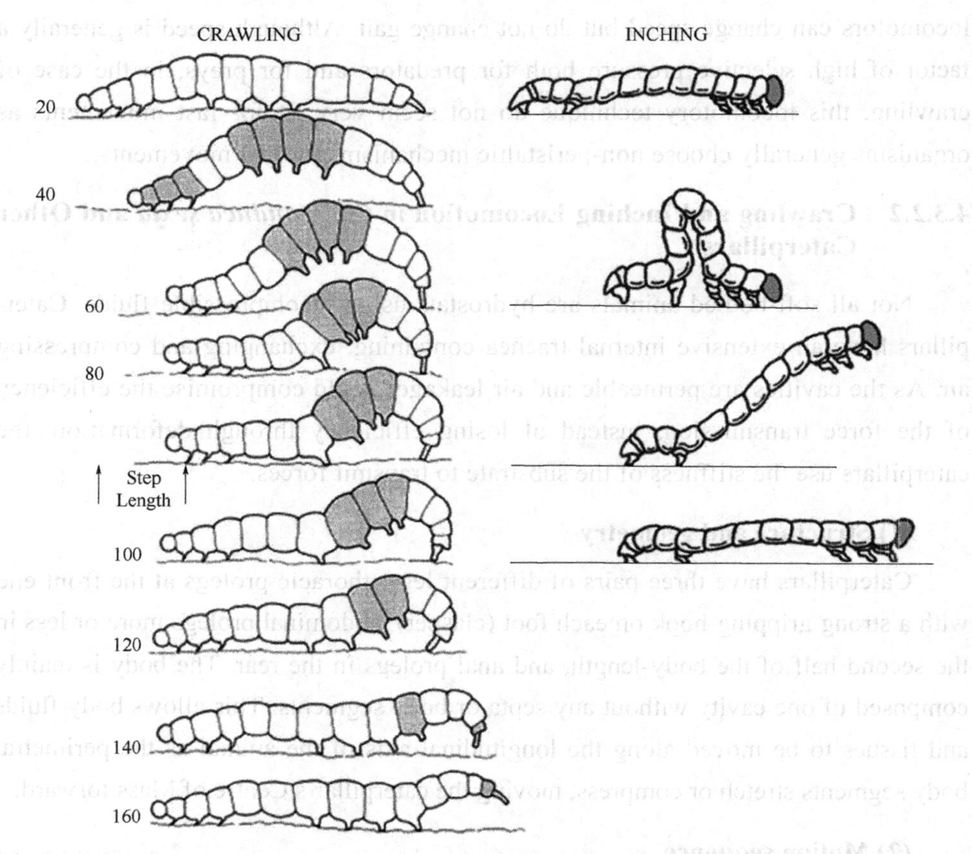

Fig.4.9 Crawling locomotion in a caterpillar, and inching locomotive sequence. In both figures, the animals move from left to right

As in vertebrates, to change speed, caterpillars modify the cycle frequency by increasing stride length, stride frequency or muscle force with a consequent increase of energy cost per stride. The typical movement patterns achieved and the body morphology constrain the animals to take short stride lengths (about 30% of that of animals with solid skeletons) that allow only locomotion at slow speeds. Faster maneuvers can however be performed by reversing the waves in several gaits, which however brings potential instability:

- Reverse walking, with inverted kinematics of the forward walking;
- Reverse galloping, where all legs but the hooks are off the ground in each step;
- Recoil and roll—a unique form of locomotion—where the caterpillar

quickly bends inwards, the front touching the back, and rolls down from its support making use of gravitational potential force.

(3) Motion mechanism

Caterpillars, as a hydrostatic system, separate from the standard constant body-volume constraints of most hydrostats for a number of reasons:

- Longitudinal extension is due to inter-segmental folds, not to stretching;
- Body pressure is variable during motion due to a fluid flow;
- Air can escape the volume through the trachea;
- The animal is not compartmentalized and cannot control pressure locally;
- The animal has leg-contact with the substrate and does not use the typical body-contact frictional-crawling type of body mass transfer;
- Mechanical behavior in the tissues is non-linear and anisotropic.

Caterpillars control their body stiffness mainly with pressurization of the body shell and muscle activation, employed simultaneously. When crawling, contact with the substratum is mainly provided through hooking the claspers to the surface. As the wave of segmental contraction moves forward, muscles are activated to lift and unhook the prolegs so they can be moved forwards and hook to the next spot.

4.3.2.3 Tube Feet, Hydrostatic Locomotion in Echinoderms

Echinoderms have a hydrostatic skeleton and move through a multitude of small hydraulic feet, which can change shape and orientation by using the displacement of a fluid inside their cavity. The feet are not only used for locomotion, but also for burrowing, feeding, breathing and sensing.

A direct biomimetic application of these structures is pneumatic artificial muscles, which is made of a flexible tubular membrane reinforced with fibers. Change of pressure in the tube causes it to extend or contract (Fig.4.10).

(1) Structure and geometry

All living echinoderms are pentaradial symmetric with an internal skeleton and a body cavity (coelom) with water-filled canals (ambulacra or water-vascular). In the sea stars in particular, the ambulacra is connected with the tube feet: hundreds of small cylinders or tentacle-like protrusions of the body wall extending in rows through holes in the skeleton. On the tip of the tube, feet are suction cups, which allow the starfishes to stick to a surface. Each cylindrical tube is connected to a muscular bulb (ampulla) aligned with the muscular bulbs of the other tube feet on the inner side of the ambulacra. The tube foot-ampulla system is connected and supplied with water through the radial canal through a one-way valve (Fig.4.11).

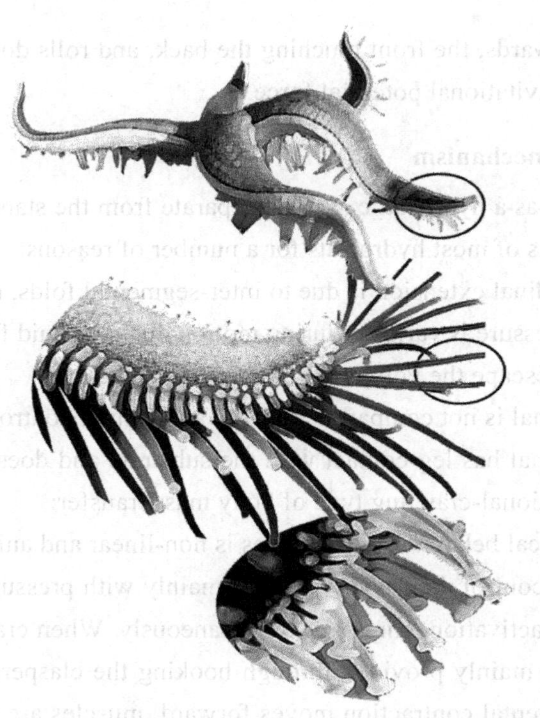

Fig.4.10 Close-up of a walking asteroid and the tube feet

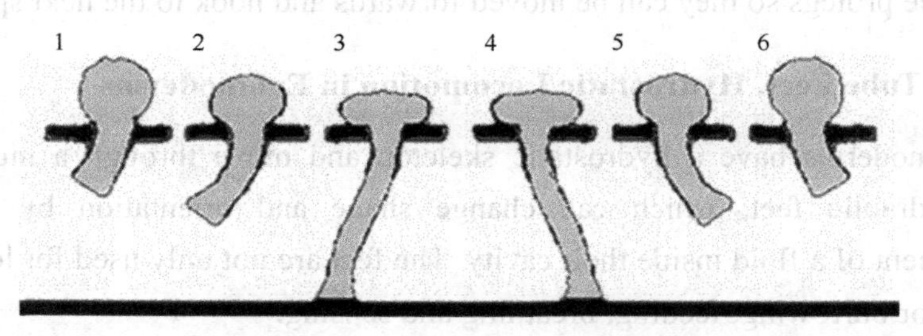

Fig.4.11 Extension cycle of the tube feet

Most important thing for the motion and directing of the tube feet is the arrangement of circular and longitudinal muscle fibers able to contract selectively, antagonizing each other. The walls of the tube feet are reinforced with cross-fiber helical connective tissues, allowing shape change through extension and bending but preventing torsion and circumferential inflation.

(2) Motion sequence

Sea stars crawl around on rocky and muddy bottoms moving forward with any side of the body and reverse direction without having to turn around. Only one arm however becomes dominant in the locomotion, while the tube feet on all other arms move in the same plane as the lead arm.

The tube feet are used as levers to move forward, or to stick to any other mechanisms that can pull it forward. They appear to move quite uncoordinated apart from small areas.

Studies made on echinoderms do not reveal if any variations in gait can be taken with this type of locomotion, but other special movements have been recorded.

- Righting response. As in most echinoderms, the starfish's normal position is with the mouthparts facing a surface, if turned upside down, it performs a slow somersault using its arms and tube feet to return to its normal position.
- Burrowing. Several asteroid species bury themselves in sand or mud (Fig.4.12).

Fig.4.12 Righting response sequence in a sea star

(3) Motion mechanism

The mechanism relies on the interaction of two connected pouches—the ampullae and the tube foot—which antagonize each other by alternatively inflating and deflating transmitting force through the incompressibility of the vascular fluid (seawater). Decrease in volume in the ampullae (through contraction) forces the fluid into the tube foot, resulting in its extension. As the muscles of the tube feet contract, the water is forced back into the ampulla which refills with the same liquid. The muscles in the wall of the tube foot control its direction and bending movements. The importance of the helical arrays is mainly due to the fibers' angle of inclination, which transmits deformation to the tube it surrounds.

4.3.3 Rigid Systems

This part shows few examples of animals using mechanics of rigid systems, vertebrates and shell-animals (endo- and exo-skeletons). The two case studies have consciously been chosen because of the relative simplicity and clarity of the specifically studied aspects, to assure an easier potential biomimetic technology transfer of the analyzed concepts.

These systems all rely on displacement through the actuation of different variations of the lever system: the bivalves use a simple lever system, opening and closing the shell; the insect wings have evolved an advanced lever system to fly at high frequencies.

The case-study sheets explore the following rigid-body structures, focusing on one of the reactive kinetic structures and mechanisms (Table 4.3).

Table 4.3 The kinetic aspect of rigid-body structures

Name of the organism	Kinetic aspect studied
Scallop	Valve clapping locomotion
Insect	Indirect flight

4.3.3.1 Scallop Valve-Clapping Swimming

Bivalves are freshwater mollusks (among others: clams, oysters, cockles, mussels, scallops, etc.), consisting of two symmetrical shells protecting the animal. Out of 17, 000 known species of bivalves, only a few are capable of swimming: scallops in particular have evolved this capacity to escape predation (Fig.4.13).

Fig.4.13 Three different types of scallop shells (top and side view)

From the left: *Amusium balloti*, *Pecten fumatus*, and *Mimachlamys asperrima*

Motion is achieved through a very basic lever system which relies on a streamlined light shell (the beam), a strong adductor muscle (the effort), and a very elastic hinge connecting the two shells (the fulcrum). By clapping the valves together, a water jet is expelled from the dorsal edge of the shell, providing thrust in the opposite direction to the opening of the valves.

(1) Structure and geometry

Scallops exhibit a very wide range of shell morphologies, from very streamlined to less hydrodynamic ones. All however rely on the same locomotion characters and body parts. The two rigid specular shells are connected through a hinge enclosing and protecting the soft-bodied mollusk, which actuates its exoskeleton through the action of one adductor muscle alone, providing contraction. A very important body part for the well functioning of the mechanism is the hinge ligament, which connects the two shell halves. This ligament has excellent elastic proprieties, acting as a flexible hinge to allow the smooth and repeated closure of the valves, and provides antagonist force to the only muscle the scallop possesses. While the outer part of the ligament holds the two valves together and in place on the hinge, the inner part is a highly elastic cushion, compressing as the shells are shut close, releasing elastic energy and extending the adductor muscle again as soon as it relaxes, opening the valves before the next contraction.

The bivalve's swimming ability is determined by factors as shell shape and the adductor muscle's metabolic capacity. Thrust is achieved through water jet propulsion; drag and lift depend on the swimming angle and speed, which are determined by the shell morphology. In scallops, good swimmers generally have light valves with smooth surfaces, and a high aspect ratio due to a slightly more convex upper valve, in comparison to the lower valve (Fig.4.14).

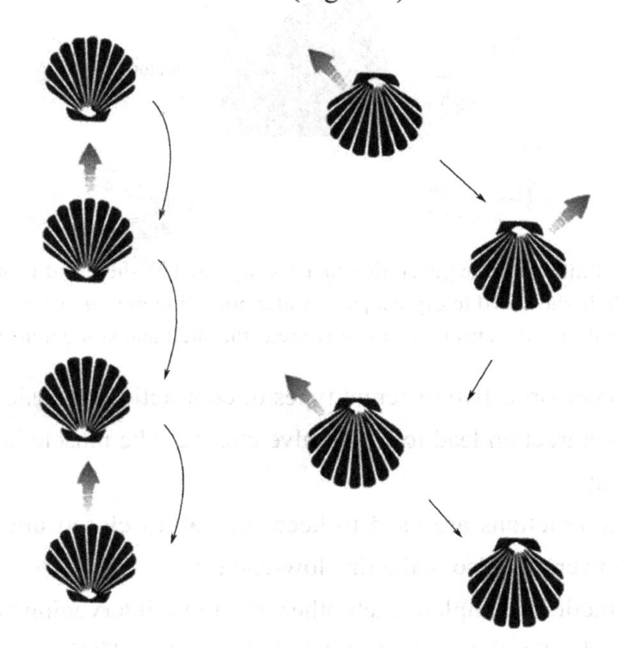

Fig.4.14 Two modes of locomotion in the scallop. The black arrows indicate the direction of locomotion (proceeding top-down), and the grey arrows indicate the direction of the expelled water jet. The locomotive pattern on the left is a unidirectional jumping motion, where the scallop moves backwards. On the right, a zig-zagging locomotion as the scallop jumps forwards alternatively to the right and to the left side

(2) Motion sequence and mechanism

The motion sequence in scallop swimming can be summarized in three steps:

- Relaxed state. The shells are wide open, the adductor muscle and the spring ligament are relaxed.

- Activation. As the scallop is disturbed, it shortens the adductor muscle to snap the shells closed, compressing the elastic ligament and producing a water jet as a consequence of the fast shell closure, which thrusts the animal through the water.
- As the muscle is relaxed, the shell reopens through passive recoil of the elastic hinge ligament.

Swimming is achieved by repeating the opening/snapping motion several times per second, with phases of gliding through the water in between. The adductor muscle in the scallop is particularly strong in proportion to its body mass, in comparison to not-swimming bivalves: the pressures created by the muscle can reach up to 2 kgf/cm^2, producing the strong water jets on which the animals depend to locomote (Fig.4.15).

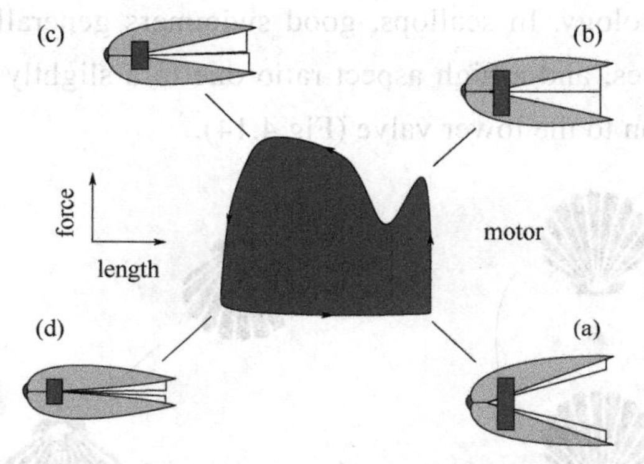

Fig.4.15 Phases of muscle force generation in a scallop. (a) The shell is at its maximal opening; (a)-(c) the muscle is shortened to drive a jet of water providing motion; (d) the shell is closed; (d)-(a) force drops, the elastic ligament reopens the shell and straightens the muscle

The scallop performs two different types of contractions: phasic and tonic.

- Phasic contraction lead to rapid valve closure. The muscle is fast and highly energy demanding.
- Tonic contractions are used to keep the valves closed under a long time. The muscle is slower, but also maintains low-energy.

These contractions complete each other: the tonic intervening when the phasic muscle is fatigued, allowing partial metabolic recovery. Timing and frequency of phasic and tonic contraction varies considerably between species, which have adopted specific combinations to better fit their morphology, habitat and lifestyle.

4.3.3.2 Insect Flight, Indirect High-Frequency Flight

Insects evolved active flight first among all animals. With multiple and complex aerodynamic maneuvers, as taking off backwards, flying sideways and landing upside down, they still remain unbeaten under many aspects. Because of their small body size (20 μg-3 g), flying insects must beat their wings at a much higher frequencies than vertebrates to generate enough aerodynamic thrust.

(1) Structure and geometry

The wings, thin membranes structurally straightened by a web of veins, are an outgrowth of the exoskeleton, enabling the insects to use a lever system to provide thrust. The wings are found on the second and third thoracic segment, and can occur in one or in two pairs (in that case described as fore and hind-wings). Wings in two pairs are mechanically linked together and work in most cases as a single wing.

With the lever system, a bigger wing area is activated through the action of the muscles on the one end of the structure. The rigid part of the wing (the beam), which mechanically transmits the muscle force to the wing surfaces, is connected to the thorax of the insect through an articulation (the fulcrum) and continues shortly inside the thorax (the effort), building a lever system similar to a seesaw. The flight muscles do not directly activate the wing beat as in vertebrates, but deform the thorax, applying the effort on the opposite end of the load, with the fulcrum in the middle.

(2) Motion sequence

There are two main models of insect flight: leading edge vortex and fling-and-clap. Other gaits are hovering and gliding.

Leading edge vortex is created through flapping the wings in two half-strokes:

- The down stroke starts up and backwards, moving forwards down;
- The wing is tilted around its axis, with the edge facing backwards, to minimize force on the upward stroke;
- The upstroke pushes the wing up and back;
- The wing is tilted again to the original position.

Fling-and-clap is used by very small insects. In this motion pattern, the wings are clapped together above the insect's body and then flung apart. During the flinging motion, negative pressure is created above the animal, so air is sucked in creating a vortex over each wing creating lift. As the vortex moves over the wing, it also contributes to the clapping movement. The reason why bigger insects do not use this motion scheme, is that the repeated clapping would cause damage and wear to the wing.

In hovering, the insect stays on the same spot in the air, not moving forward. In this pattern, the wing beats twice as fast as in forward flight, requiring lift as well as stabilization. As the insect starts flying forwards and flight speed increases, it also tilts its body and head down, minimizing the body area in contact with the airflow and the drag (Fig.4.16).

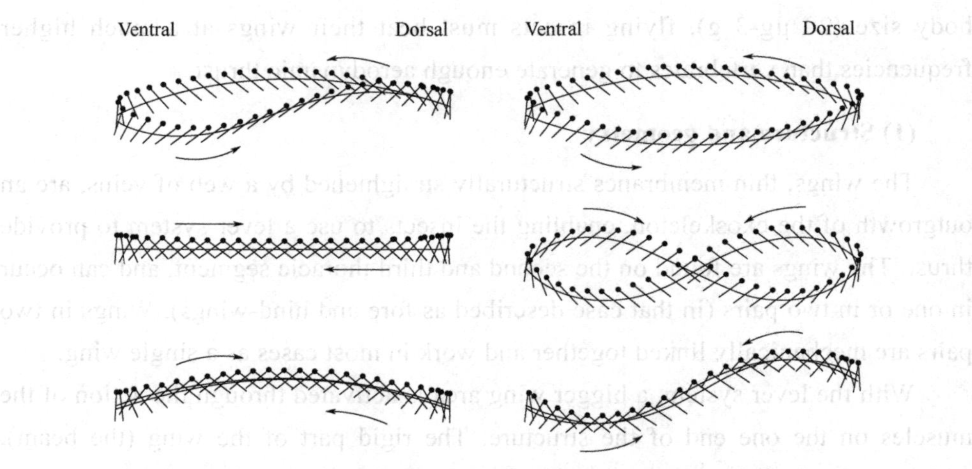

Fig.4.16 Six possible variations of wing stroke patterns in insects

Flies with two pairs of wings have evolved the two hind wings (halteres) to oscillate anti-phase to the front wings. These help the insects to identify rotations of the body during flight, and are therefore important for aerial maneuvers. The mechanism underlying the synchronization between front and hind wings is not yet well understood, but seems to be due to specific mechanic linkages within the thorax of the insect.

(3) Motion mechanism

The thorax works as a rigid box to which the wings are attached. As previously discussed, movement in the thorax brings movement in the wing thanks to two pairs of muscles:

- The first muscle set runs parallel to the dorso-ventral axis, connecting the thorax to the back of the insect. Contraction depresses the thorax area and raises the wing due to a reverse linkage between thorax and the wing.
- The other set of muscles run diagonally from the anterior of the thorax to the posterior floor. Contraction in the diagonal muscles elevates the thorax, extends the first set of muscles and provides the down stroke (Fig.4.17).

Fig.4.17 Actuation of the dorso-ventral muscles (left) and the tergosternal muscles (right), deforming the thorax and aiding the indirect motion of the wings

In a few species, the down stoke is provided by an elastic recoil of the thorax as the muscles relax. Also, at the end of each down stroke, the wing needs to be slowed down to reverse for the upstroke. During down stroke, the muscles dissipate energy by converting it into heat, or (in a few species) converting it into potential energy for the upstroke through the action of an elastic pad (resilin). As the wing beats the upstroke, the resilin is stretched, giving back energy into the down stroke by snapping back.

Because nervous systems cannot control stroke-by-stroke wing beats at such speeds, some insects (as Neoptera) have evolved a unique feature: the asynchronous flight muscle. This muscle is able to beat at higher frequencies (over 1,000 Hz) than the typical flight muscle (5-200 Hz). A single nerve impulse causes the muscle fiber to contract multiple times, allowing the frequency produced to exceed by many times the rate at which a nervous system is able to cope with normal wing beat impulses.

Key vocabularies

abdomen [ˈæbdəmən] ***n.*** 腹部；下腹；腹腔
abdominal [æbˈdɒmɪnl] ***adj.*** 腹部的;有腹鳍的
actuation [ˌæktʃʊˈeʃən] ***n.*** 刺激，冲动；吸合
adductor [əˈdʌktə(r)] ***n.*** 内收肌；内转肌
aesthetics [iːsˈθetɪks] ***n.*** 美学；美学理论；艺术美
agility [əˈdʒɪlətɪ] ***n.*** 敏捷，灵活；机敏
agonist [ˈægənɪst] ***n.*** 主动肌；兴奋剂，激动剂；促效药
airfoil [ˈeəfɔɪl] ***n.*** 机翼；螺旋桨；翼面
ajar [əˈdʒɑː(r)] ***adj.*** 半开的；微开的；不和谐的. ***adv.*** 半开地；不协调地；不和谐地
ambulacra [æmbjʊˈleɪkrə] ***n.*** ambulacrum 的名词复数：柄吸盘，吸垫，足吸盘
ampulla [æmˈpʊlə] ***n.*** 壶状体；壶腹
anal [ˈeɪnl] ***adj.*** 肛门的；直肠的；肛门附近的
anatomy [əˈnætəmi] ***n.*** 解剖学；组成结构，解剖构造；人体；剖析；骨骼
anchor [ˈæŋkə(r)] ***v.*** 使稳固，使固定；成为支柱；使扎根，使基于；支持，保护
angiosperm [ˌændʒɪəˌspɜːm] ***n.*** 被子植物
antagonist [ænˈtægənɪst] ***n.*** 对手，敌手，对立者；拮抗物；拮抗肌
antagonize [ænˈtægənaɪz] ***vt.*** 使成为敌人；引起……敌对；对……起反作用；中和
anterior [ænˈtɪəriə(r)] ***adj.*** 位于前部的；先前的；早期的；在前的

anterograde [æntəˈrɒgreɪd]***adj.*** 顺行的；前进的
apex [ˈeɪpeks] ***n.*** 顶点，最高点；茅苗的顶端
appendage [əˈpendɪdʒ] ***n.*** 附属物；附属器官；附属肢体
ascension [əˈsenʃn] ***n.*** 上升，升高
aspect ratio ***n.*** 纵横比；屏幕高宽比；宽高比
asteroid [ˈæstərɔɪd] ***n.*** 海星. ***adj.*** 星状的；类似海星的
asynchronous [eɪˈsɪŋkrənəs] ***adj.*** 异步的；不同时的；不同期的
bipinnate [bɪˈpɪneɪt] ***adj.*** 二回羽状的，两羽状的
bivalve [ˈbaɪvælv]***n.*** 双壳贝；双壳类动物. ***adj.*** 双壳的；两瓣的
bulb [bʌlb] ***n.*** 鳞茎，球茎；球状物
burrow [ˈbʌrəʊ] ***v.*** 掘地洞，挖地道；搜寻；挖穿
cantilever [ˈkæntɪliːvə(r)] ***n.*** 悬臂
carnivorous [kɑːˈnɪvərəs] ***adj.*** 食肉的
cilia [ˈsɪlɪə] ***n.*** 睫毛、纤毛（cilium 的名词复数）
circumferentially [səˌkʌmfəˈrenʃəli] ***adv.*** 圆周地，环绕地；迂回地，间接地
clam [klæm] ***n.*** 蚌，蛤；钳子
clasper [kˈlɑːspər] ***n.*** 扣，卷须，尾脚；抱握器
cockle [ˈkɒkl] ***n.*** 海扇壳，海扇类；鸟蛤
coelom [ˈsiːləm] ***n.*** 体腔
compartmentalize [ˌkɒmpɑːtˈmentəlaɪz] ***vt.*** 划分，区分
concave [kɒnˈkeɪv] ***adj.*** 凹的，凹面的. ***n.*** 凹面. ***v.*** 使变凹形
concertina [ˌkɒnsəˈtiːnə] ***v.***（像六角手风琴一样）折起，折叠
convex [ˈkɒnveks] ***adj.*** 凸的，凸面的
corolla [kəˈrɒlə] ***n.*** 花冠
curvature [ˈkɜːvətʃə(r)] ***n.*** 弯曲；弯曲部分；曲率；曲度
descension [dɪˈsenʃən] ***n.*** 降下；祖先
diagonal [daɪˈægənl] ***adj.*** 对角线的；斜的；沿斜线运动的. ***n.*** 斜线，斜纹；对角线
Dionaea [ˌdaɪəˈniːə] ***n.*** 捕蝇草属
dissemination [dɪˌsemɪˈneɪʃn] ***n.*** 宣传，散播
dorsal [ˈdɔːsl] ***adj.*** 背部的；后背的；远轴的
dorso-ventral ***adj.*** 后前位的，背腹侧的
echinoderm [eˈkaɪnədɜːm] ***n.*** 棘皮动物
encumbrance [ɪnˈkʌmbrəns] ***n.*** 负担；累赘；妨碍；阻碍
endoskeleton [ˈendəʊskelɪtn] ***n.*** 内骨骼
enzyme [ˈenzaɪm] ***n.*** 酶
epidermal [ˌepɪˈdɜːməl] ***adj.*** 表皮的，外皮的

exoskeleton [ˈeksəʊskelɪtn] ***n.*** 外骨骼

exploitation [ˌeksplɔɪˈteɪʃn] ***n.*** 剥削，压榨；开发，开采

fine-tuned ***vt.*** 调整；使有规则；对进行微调

fling [flɪŋ] ***vt. & vi.*** 猛扑；猛冲；急伸． ***n.*** 投掷

footstalk [ˈfʊtstɔːk] ***n.*** 叶柄；花梗；总柄；轴柄

fulcrum [ˈfʊlkrəm] ***n.*** 支点；叶附属物

funnel [ˈfʌnl] ***n.*** 漏斗；烟囱；漏斗状物；通风井

furl [fɜːl] ***v.*** 叠，卷收；收拢，卷起． ***n.*** 卷起，折叠；卷起之物

gait [geɪt] ***n.*** 步态，步法；花纹循环；速度

gallop [ˈgæləp] ***v.*** （马）飞跑，飞奔；骑马奔驰，飞速移动

geotropism [dʒiːˈɒtrəpɪzəm] ***n.*** 向地性；屈地性

glabrous [ˈgleɪbrəs] ***adj.*** 无毛的；光洁的

haltere [ˈhɔːltɪə(r)] ***n.*** 缰绳；笼头；平衡棒

haptotropism [hæptəʊˈtrəʊpɪzəm] ***n.*** 向触性，趋触性

helically [ˈhelɪkəlɪ] ***adv.*** 成螺旋形地

heliotropism [ˌhiːlɪˈɒtrəpɪzəm] ***n.*** 向日性，趋日性

helix [ˈhiːlɪks] ***n.*** 螺旋，螺旋状物；耳轮

herbivore [ˈhɜːbɪvɔː(r)] ***n.*** 食草动物

hydrostat [ˈhaɪdrəstæt] ***n.*** 汽锅爆裂防止器；水压调节器；电位水位指示器

hydrostatic [ˌhaɪdrəˈstætɪk] ***adj.*** 静水力学的，流体静力学的

hydrotropism [haɪˈdrɒtrəpɪzəm] ***n.*** 向水性

inching [ˈɪntʃɪŋ] ***n.*** 点动；微调尺寸；寸动；缓动

interven [ˌɪntəˈviːn] ***v.*** 出面；介入；阻碍；干扰；插嘴；介于……之间

Ipomoea [ˌɪpəˈmiːə] ***n.*** 番薯属植物（圆叶牵牛，紫花牵牛等）

juxtapose [ˌdʒʌkstəˈpəʊz] ***vt.*** 把……并列，把……并置，把……放在另一个旁边

kinematics [ˌkɪnəˈmætɪks]***n.*** 运动学；动力学

kinetic [kɪˈnetɪk] ***adj.*** 运动的，活跃的，能动的，有力的；动力（学）的，运动的

lamina [ˈlæmənə] ***n.*** 薄板；薄层；叶片

leaflet [ˈliːflət] ***n.*** 传单，小册子；小叶，初生叶（统称）． ***v.*** 散发传单（或小册子）

ligament [ˈlɪgəmənt] ***n*** 韧带；纽带

limb [lɪm] ***n.*** 肢；大树枝；（某物的）突出部，延伸部分；（管状花的）瓣片

lobe [ləʊb] ***n.***（脑、肺等的）叶；裂片；耳垂；波瓣

Manduca sexta ***n.*** 烟草天蛾

metabolic [ˌmetəˈbɒlɪk] ***adj.*** 新陈代谢的；变化的

Mimosa pudica ***n.*** 含羞草

mollusk [ˈmɒləsk] ***n.*** 软体动物
morning glory ***n.*** 牵牛花（番薯属植物）；昙花一现的人或物
morphological [ˌmɔ:fəˈlɒdʒɪkl] ***adj.*** 形态学的，形态的
mussel [ˈmʌsl] ***n.*** 贻贝，蚌类；淡菜
nastic [ˈnæstɪk] ***adj.*** 倾性的，感性运动的
Neoptera [niˈɔptərə] ***n.*** 新翅目；新翅下纲
nocturnal [nɒkˈtɜ:nl] ***adj.*** 夜间的；夜间活动的；夜间开花的
nutritive [ˈnju:trɪtɪv] ***adj.*** 营养的. ***n.*** 富于营养的食物
nyctinastic [nɪktɪˈnæstɪk] ***adj.*** 感夜的
oblong [ˈɒblɒŋ] ***adj.*** 长方形的；矩形的；椭圆形的；椭圆体的
oscillate [ˈɒsɪleɪt] ***v.*** 振荡；振动；动摇
outpace [ˌaʊtˈpeɪs] ***vt.*** 超过……速度，赶过
oyster [ˈɔɪstə(r)] ***n.*** 牡蛎，蚝；牡蛎白
palmated [ˈpælmeɪtɪd] ***adj.*** 掌状的；有蹼的
Passiflora ***n.*** 西番莲属
pentameric [pentæˈmerɪk] ***adj.*** 五聚物的
peristalsis [ˌperɪˈstælsɪs] ***n.*** 蠕动
peristaltic [pəˈrɪstæltɪk] ***adj.*** 蠕动的
permeable [ˈpɜ:miəbl] ***adj.*** 可渗透的，具渗透性的
pertain [pəˈteɪn] ***v.*** 适合；关于；适用；生效；存在
petiole [ˈpetɪəʊl] ***n.*** 叶柄，柄部
phasic [ˈfeɪzɪk] ***adj.*** 形势的；相位的；阶段的
photonastic ***adj.*** 感光的
phototropism [ˌfəʊtəʊˈtrəʊpɪzəm] ***n.*** 趋光性，向光性
phytochrome [ˈfaɪtəʊkrəʊm] ***n.*** 光敏色素；植物色素
pigment [ˈpɪgmənt] ***n.*** 色素；颜料. ***v.*** 给……染色；呈现颜色
pneumatic [nju:ˈmætɪk] ***adj.*** 气动的；充气的；有气胎的. ***n.*** 气胎
pollinator [ˈpɒlɪneɪtə] ***n.*** 传粉者，传粉媒介，传粉昆虫；授花粉器
predation [prɪˈdeɪʃn] ***n.*** 捕食；掠夺
prostrate [ˈprɒstreɪt] ***adj.*** 卧倒的，匍匐的，平卧的
protrusion [prəˈtru:ʒn] ***n.*** 突出，伸出；突出物
pulsatile [ˈpʌlsətaɪl] ***adj.*** 脉动的；跳动的；悸动的
pulvinus [pʌlˈvaɪnəs] ***n.*** 叶枕；呈鼓突状
recoil [rɪˈkɔɪl, ˈri:kɔɪl] ***vi.*** 反跳，跳回；后退. ***n.*** 后退，退缩；反跳，跳回；反冲力
resemble [rɪˈzembl] ***v.*** 像，与……相似
resilin [ˈrezɪlɪn] ***n.*** 节肢弹性蛋白
retrograde [ˈretrəgreɪd] ***adj.*** 倒退的；退化的；次序颠倒的

rub [rʌb] ***v.*** 擦，摩擦；揉擦；磨损
scallop [ˈskɒləp] ***n.*** 扇贝；扇贝壳；扇（贝）形．***vt.*** 使成扇形．***vi.*** 拾扇贝
seismonastic [saɪzməʊˈnæstɪk] ***adj.*** 感震（性）的
septa [ˈseptə] ***n.*** 隔膜（septum 的复数）
serpentine [ˈsɜːpəntaɪn] ***adj.*** 蜿蜒的，弯弯曲曲的；复杂的
sluice [sluːs] ***n.*** 水闸；有闸人工水道
snap [snæp] ***v.*** 断裂，折断；打开
somersault [ˈsʌməsɔːlt] ***n.*** 翻筋斗
specular [ˈspekjʊlə] ***adj.*** 镜的；窥器的；用窥器（检查）的；镜子似的；会反射的
sporangium [spɒˈrændʒɪəm] ***n.*** 孢子囊，芽孢囊
stimulus [ˈstɪmjələs] ***n.*** 刺激（物），促进因素；刺激性
stride [straɪd] ***n.*** 大步，阔步；步态，步伐；步距，步幅；步速．***v.*** 大步走，阔步走
substratum [ˈsʌbstrɑːtəm] ***n.*** 基础；根据；下层
synchronization [ˌsɪŋkrənaɪˈzeiʃn] ***n.*** 同一时刻；同步；使时间互相一致；同时性
synchronize [ˈsɪŋkrənaɪz] ***v.*** 同步，同时，对准；相符；同步化，共同行动
synergic [sɪˈnɜːdʒɪk] ***adj.*** 合作的，协作的
tendril [ˈtendrəl] ***n.*** 卷须；蔓；卷须状物
tentacle [ˈtentəkl]***n.*** 触手；触角；触须
tergosternal [tɜːgəˈstɜːnəl] ***adj.*** 背腹板的
thermotropism [θə(ː)ˈmɒtrəpɪzəm] ***n.*** 向热性，向温性
thigmotropism [θɪgˈmɒtrəpɪzəm] ***n.*** 向触性
thoracic [θɔːˈræsɪk] ***adj.*** 胸的；胸廓的
thorax [ˈθɔːræks] ***n.*** 胸，胸膛；胸腔
thrust [θrʌst] ***v.*** 猛推，猛塞；刺，扎；（人）推进，挤过；上伸，伸出
tonic [ˈtɒnɪk] ***n.*** 补药；主调音或基音．***adj.*** 滋补的；声调的；使精神振作的
tonic contraction ***n.*** 强直性收缩；紧张性收缩
toroid [ˈtɔːrɔɪd] ***n.*** 环形室圆环面；螺旋管．***adj.*** 圆的，圆突的
toroidal [ˈtɔːrɔɪdl] ***adj.*** 环形的；螺旋管形的 [数学]，超环面的．***n.*** 曲面（圆环）
trachea [trəˈkiːə] ***n.*** 气管；导管
tropism [ˈtrəʊpɪzəm] ***n.*** 取向；向性；趋性；向性运动
turgid [ˈtɜːdʒɪd] ***adj.*** 肿胀的；浮夸的；浮肿的
turgor [ˈtɜːgə] ***n.*** 细胞（组织）的膨胀；肿胀；浮夸；夸张
unroll [ʌnˈrəʊl] ***vt.*** 展开，铺开；显示．***vi.*** 显露；展现
upstroke [ˈʌpstrəʊk] ***n.*** 向上的一击；书写时向上的一笔；上行程；上行运动
vascular [ˈvæskjələ(r)] ***adj.*** 脉管的；血管的；含有血管的；充满活力的
vectorial [vekˈtɔːrɪəl] ***adj.*** 媒介物的；向量的；矢的；矢量的；带菌体的

Venus flytrap ***n.*** 捕蝇草
vertebrate [ˈvɜːtɪbrət] ***n.*** 脊椎动物. ***adj.*** 有脊椎的，脊椎动物的
zig-zag ***adj.*** 锯齿状的，曲折的

Exercises

Ⅰ. Translate the following English into Chinese.

1. Adaptivities on a micro- and macro-scale are however interrelated and complementary. In nature's constructions, form and function are integrated in a same system, avoiding maximization of individual functions (which might harm other functions), rather optimizing the whole. There is an extremely refined combination of high-tech and low-tech solutions in laws of nature, complementing each other.

2. Moving components accommodate specific ranges of activities by changing geometry or orientation. On one hand, this gives the objects their function and usability; on the other, it makes them subject to use and wear. When motion is made impossible, the usability of the object declines and the article becomes waste, hence the need of integrating adaptivity through disassembly together with adaptivity through motion.

3. Tropism indicates a slow adaptive motion of plants, generally achieved through growth, in response to different environmental triggers where the direction of the motion is strongly dependent on the direction of the stimulus. Nastic movements are faster and differ from tropic movements as they concern non-directional movement in response to a stimulus.

4. The medium of locomotion strongly defines the physical characters of the animal. All animals are perfectly adapted to the medium they move in, and many have evolved the capacity to do so in more than one medium: frogs for instance, have evolved limbs for terrestrial locomotion and webbed feet for aquatic locomotion.

5. Earthworms use their internal body liquid to amplify the force, velocity and kinetic effect of their muscle-contraction by regulating the pressure in their body segments. As the circumferential muscles of a segment contract while the amount of fluid it contains remains constant, the cylinder is elongated from thick to thin.

Ⅱ. Answer the following questions based on the main text in this chapter.

1. What are the time-related forms for movements or change in nature? Which timeframe is appropriate in the context of designing motion using biomimics methods?

2. What are the three main ways to control a kinetic transition process in a

predictable and safe manner in biological organisms?

3. How can the plants make movements? What are the major types of movements in plants?

4. For caterpillar locomotion, what is the motion sequence and how does a caterpillar adapt the technique?

5. Why do some insects have the asynchronous flight muscle?

Ⅲ. Writing.

Search any interesting movements of plants or animals beyond the cases mentioned in this chapter and give an introduction or explanation. How can the mechanism or move sequence be abstracted to use for bionic design?

Chapter 5

Bioinspired Materials

Natural structural materials are built at ambient temperature from a fairly limited selection of components. They usually comprise hard and soft phases arranged in complex hierarchical architectures, with characteristic dimensions spanning from the nanoscale to the macroscale. The resulting materials are lightweight and often display unique combinations of strength and toughness, but have proven difficult to mimic synthetically. Here, we review the common design motifs of a range of natural structural materials, and discuss the difficulties associated with the design and fabrication of synthetic structures that mimic the structural and mechanical characteristics of their natural counterparts.

5.1 Background

The technological development of humanity was supported in its early stages by natural materials such as bone, wood and shells. As history advanced, these materials were slowly replaced by synthetic compounds that offered improved performance. Today, scientists and engineers continue to be fascinated by the distinctive qualities of the elegant and complex architectures of natural structures, which can be lightweight and offer combinations of mechanical properties that often surpass those of their components by orders of magnitude. Contemporary characterization and modelling tools now allow us to begin deciphering the intricate interplay of mechanisms acting at different scales (from the atomic to the macroscopic) and that endow natural structures with their unique properties. At present, there is a pressing need for new lightweight structural materials that are able to support more efficient technologies that serve a variety of strategic fields, such as transportation, buildings, and energy storage and conversion. To address this challenge, yet-to-be-developed materials that would offer unprecedented combinations of stiffness, strength and toughness at low density, would need to be fashioned into bulk complex shapes and manufactured at high volume and low cost. It is an open question that how this goal can be achieved. Although remarkable examples have arisen from the laboratory, it remains uncertain whether they can be scaled-up for use in practical applications.

It is a classic materials-design problem that the two key structural properties—strength and toughness—tend to be mutually exclusive (Box 1); strong materials are invariably brittle, whereas tough materials are frequently weak. Here is where natural organisms provide a rich source of inspiration for fresh ideas. They provide an opportunity for us to benefit from the great number and considerable diversity of solutions, perfected over millions of years of evolution. For example, highly mineralized, mostly ceramic, natural structures, such as tooth enamel or nacre, minimize wear and provide protection. A unique aspect of these materials is that they utilize different structures or structural orientations, to generate hard surface layers so as to resist wear and/or penetration, and have a tough subsurface to accommodate the increased deformation; that is, unlike human-made hard materials such as ceramics, they are designed for total fracture resistance. Specifically, they arrest crack propagation and avoid catastrophic failure. Other examples of evolution-driven strategies are the use of highly porous architectures in materials that must combine light weight and stiffness, such as cancellous bone or bamboo. However, stiff and porous structures tend to be weak. To maintain strength, some natural materials feature complex designs that frequently incorporate nanofibres and intricate architectural gradients. In contrast to most human-engineered materials, such natural materials are built at ambient temperatures through bottom-up strategies that are difficult to duplicate in large-scale manufacturing. Most importantly, natural materials that combine the desirable properties of their components often perform significantly better than the sum of their parts—an advantage that has sparked much of the current interest in bioinspired materials design. In particular, they offer a path towards solving the challenge of designing materials that are both strong and tough, through the development of a confluence of mechanisms that interact at multiple length scales, from the molecular to the macroscopic. Hybrid materials that are highly mineralized (such as seashells, bone and teeth), lightly mineralized (such as fish scales and lobster cuticle) and purely polymeric (such as insect cuticle, wood, bamboo or silk) are prime examples of natural composites with properties that far exceed those of their material constituents.

Mimicking the features of a natural material is not a trivial undertaking. Many investigators have characterized the nano/microstructure of a wide variety of natural structural materials—ranging from wood, antler, bone and teeth, to silk, fish scales, bird beaks and shells. Yet few have comprehensively characterized the most critical mechanical properties of these materials, such as strength and toughness. Fewer still have identified the salient nano/microscale mechanisms underlying such properties. Equally important, there are even fewer examples so far of practical synthetic versions of these materials. Consequently, critics have suggested that the field has been largely unsuccessful in its quest to apply bioinspired strategies to engineering materials and design. However, as exemplified by the development of nacre-inspired materials, the first steps have been taken in characterization, modelling and manufacturing. Such progress is fueling the growing conviction that highly damage-tolerant bioinspired structures can be designed and built.

Because natural materials typically feature a limited number of components that have relatively poor intrinsic properties, superior traits stem from naturally complex architectures that encompass multiple length scales. In contrast, most engineered materials have been developed through the formulation and synthesis of new compounds, and with structural control primarily at the micrometre scale. Consequently, it has been claimed that nanotechnology is opening a spectrum of possibilities by allowing the manipulation of materials at previously unattainable dimensions. However, with respect to complex mechanical properties such as fracture toughness—for which characteristic length scales are orders of magnitude larger, an exclusive focus on the nanoscale would be far too limited. Any rational strategy must incorporate nano-, micro- and macroscale features, and thus involve the so-called mesoscale approach.

To accomplish this, one must extract the key design parameters from natural structures—that is, their natural design motifs (Box 2)—and translate them to other material combinations. Still, it is important to keep in mind that modern engineering demands that any bioinspired process must be scaled-up for practical manufacturing so as to accelerate fabrication and reduce the time between design and implementation. In fact, in this review we contemplate whether the biomimetic approach for the creation of better structural materials will ultimately succeed. Our examination of this issue begins with a brief review of important natural structural materials and the mechanisms underlying their mechanical behavior and function, and is followed by a detailed discussion of the key lessons offered by these materials and of the difficulties encountered in attempts to implement them in practical synthetic structures.

5.1.1 Box 1 | Essentials of mechanical properties

The mechanical properties of materials describe their ability to withstand applied loads and displacements. The fundamental relationship underlying these properties is the constitutive law, which relates the strain (normalized relative displacement) that a material experiences to an applied stress (load normalized by area). This relationship can be defined to embrace many modes of deformation behavior, such as elasticity (reversible), plasticity (permanent) or rate-dependent deformation (for example, viscoelasticity or high-temperature creep), and in principle can be established at any length scale. However, it is generally measured using a uniaxial tensile test, whereby a sample is loaded in tension (or compression), and the (normal) strains are measured as a function of the applied (normal) stress to determine properties such as stiffness, strength, ductility and toughness.

Stiffness is related to the elastic modulus, and defines the force required to produce elastic deformation; as such, Young's modulus E is defined by the initial slope of the uniaxial stress-strain curve, where the strains are recoverable (elastic). Strength, defined by the yield stress at the onset of permanent (plastic) deformation or by the maximum strength at the peak load before fracture, is a measure of the force per unit area that the material can withstand. Hardness—another measure of strength—is estimated from the extent of penetration of an indenter into the surface of the material under an applied load. Ductility is a measure of the maximum strain before fracture, and is generally assessed as the per cent elongation of the sample or its relative change in cross-sectional area. Toughness measures resistance to fracture; it can be assessed in terms of the area under the load-displacement curve, but is better evaluated using the methodologies of fracture mechanics (see below).

(1) Extrinsic versus intrinsic toughening

The attainment of both strength and toughness is a requirement for most structural materials; unfortunately, these properties are generally mutually exclusive. Although the quest for stronger materials continues, they have little utility as bulk structural materials if they do not exhibit appropriate fracture resistance. It is materials with lower strength—and hence higher toughness—that is used in the most safety-critical applications, where failure is unacceptable. The development of such damage-tolerant materials has traditionally been a compromise between hardness and ductility, although there are alternative approaches based on the concept of extrinsic versus intrinsic toughening.

Lower-strength (ductile) materials develop toughness from the energy involved in plastic deformation. However, this cannot be used for brittle materials, which display little to no plasticity. To toughen these materials, one must consider fracture as a mutual competition between intrinsic damage processes, which operate ahead of a crack tip to promote its propagation, and extrinsic crack-tip shielding mechanisms, which act mostly behind the crack tip to inhibit its propagation (Fig.5.1). Intrinsic toughening acts to inhibit damage mechanisms, such as cracking or

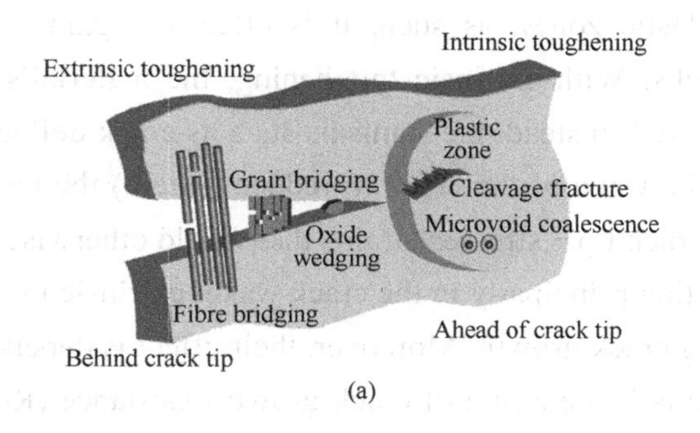

(a)

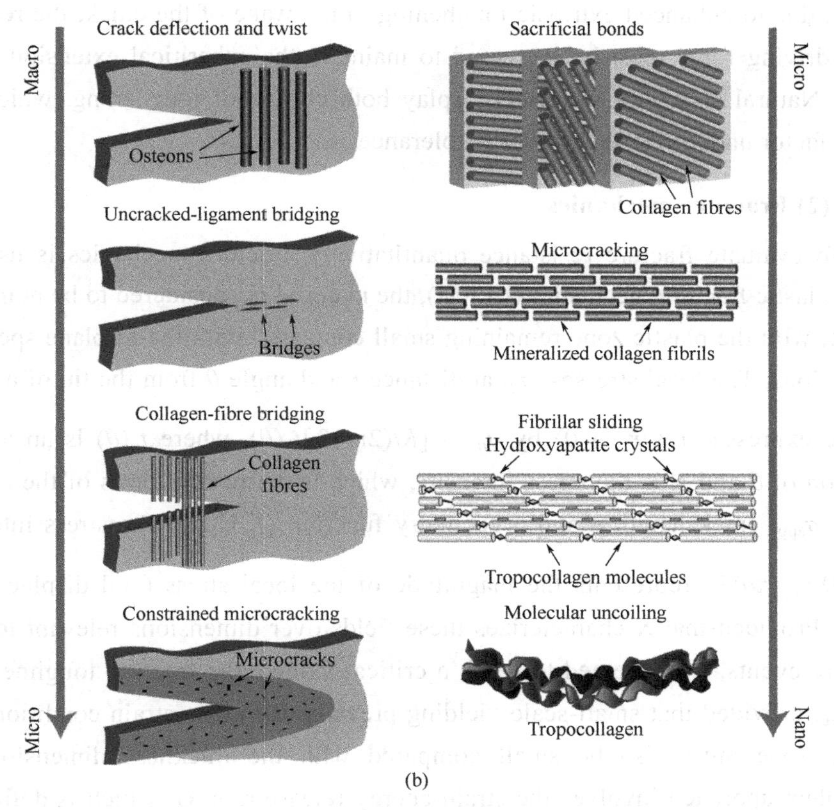

(b)

Fig.5.1 Healthy human cortical bone resists fracture through complementary intrinsic and extrinsic contributions throughout its hierarchical structure. (a) Intrinsic toughening mechanisms that promote plasticity occur ahead of the crack tip and act primarily at the nanoscale, whereas extrinsic toughening mechanisms, specifically those that shield local stresses or strains from promoting fracture, act at larger length scales and mostly behind the crack tip. (b) Intrinsically, collagen fibrillar sliding is the prime plasticity mechanism in bone, and as such has the largest impact on the inherent resistance of the hydroxyapatite/collagen composite. Other mechanisms include molecular uncoiling, microcracking and sacrificial bonding, all of which operate at submicrometre length scales. Conversely, extrinsic mechanisms, such as uncracked-ligament bridging and crack deflection, occur at micrometre length scales once the crack begins to grow so as to shield the crack tip

debonding processes, and is primarily associated with plasticity (that is, the enlarging of the plastic zone); as such, it is effective against the initiation and propagation of cracks. With extrinsic toughening, the material's inherent fracture resistance is unchanged. Instead, mechanisms such as crack deflection and bridging act principally on the wake of the crack to reduce (shield) the local stresses/strains experienced at the crack tip—stresses/strains that would otherwise be used to extend the crack. By operating principally in the crack wake, extrinsic mechanisms are only effective in resisting crack growth. Moreover, their effect is dependent on crack size. A consequence of this is the rising of crack-growth-resistance (R-curve) behaviour, where, due to enhanced extrinsic toughening in the wake of the crack, the required crack-driving force must be increased to maintain the subcritical extension of the crack. Natural structural materials display both classes of toughening, which is a major factor underlying their damage tolerance.

(2) Fracture mechanics

To evaluate fracture resistance quantitatively, fracture mechanics is used. In linear elastic fracture mechanics (LEFM), the material is considered to be nominally elastic, with the plastic zone remaining small compared with the in-plane specimen dimensions. The local stresses, σ_{ij}, at distance r and angle θ from the tip of a crack, can be expressed (as $r \to 0$) by $\sigma_{ij} \to [K/(2\pi r)^{\frac{1}{2}}]f_{ij}(\theta)$, where $f_{ij}(\theta)$ is an angular function of θ and K is the stress intensity, which is defined in terms of the applied stress, σ_{app}, crack length a and a geometry function Q. Hence the stress intensity, $K = Q\sigma_{\text{app}}(\pi a)^{\frac{1}{2}}$, represents the magnitude of the local stress (and displacement) fields. Provided that K characterizes these fields over dimensions relevant to local fracture events, it is deemed to reach a critical value—the fracture toughness—at $K = K_c$, provided that small-scale yielding prevails; for plane-strain conditions, the plastic zone must also be small compared with the thickness dimension. An equivalent approach involves the strain-energy release rate, G, which is defined as the rate of change in potential energy per unit increase in crack area. For linear elastic materials under mode Ⅰ (tensile opening) conditions, G and K are simply related by $G = K^2/E$.

LEFM-based measurements of toughness do not incorporate contributions from plastic deformation. Although many biological materials contain hard phases (for example, hydroxyapatite in bone or aragonite in nacre) that satisfy LEFM, they also comprise ductile or soft phases (such as collagen) that are the source of plasticity. When the extent of local plasticity is no longer small compared with the specimen dimensions, nonlinear elastic fracture mechanics (NLEFM) must be applied, whereby the crack-tip stress/strain fields are evaluated within the plastic (nonlinear elastic) zone. The field parameter J characterizes the local stresses/strains over dimensions comparable to the scale of local fracture events; the fracture toughness can be defined at the onset of fracture at $J = J_c$, where J is the nonlinear elastic equivalent of G in LEFM. Because of the equivalence of J and G, and in turn G and K, NLEFM enables the use of undersized specimens—too small to satisfy the stringent LEFM requirements—for measuring fracture toughness.

These toughness measurements are single-valued and pertain to where the initiation of cracking is synonymous with crack instability. In ductile materials, in many brittle materials toughened extrinsically and in most natural materials, fracture instability takes place well after crack initiation owing to the occurrence of subcritical cracking. To evaluate such crack-growth toughness, the R-curve can be used through the measurement of the crack-driving force (K, J or G) as a function of crack extension, Δa.

5.1.2 Box 2 | Common design motifs of natural structural materials

Many natural materials must be equally light, strong, flexible and tough. Because such materials are built with a relatively limited number of components, it is not surprising that we can find common design themes among them.

Natural materials often combine stiff and soft components in hierarchical structures, as is the case for nacre, bone and silk. In many of these materials, the controlled unravelling of the soft phase during fracture acts as a toughening mechanism. It therefore seems that nature's hierarchical design approach is an effective path towards combining high strength and toughness. In contrast, man-made structural composites are still far from achieving the same degree of architectural control. For example, in mineralized natural structures (such as nacre, bone or enamel), the ceramic phase is often in the form of nanometre grains, nano-platelets or nanofibres, all of which increase flaw tolerance and strength. However, in synthetic ceramic nanocomposites (with the exception of nanozirconia-reinforced ceramics), the increase in strength is not usually accompanied by a significant increase in fracture resistance.

Structural materials found in nature use carefully engineered interfaces. At the nanometre level, the chemistry of the organic component is often engineered to template the nucleation and growth of the mineral phase. Despite recent advances in the mineralization of materials in the laboratory, we are still far from effectively using mineralization as a practical technique for the large-scale fabrication of bulk structural composites. In addition, interfaces in natural materials are also designed to avoid catastrophic failure at a large scale. Whereas in the laboratory, the focus has been mostly on chemistry as a way to enhance interfacial adhesion, natural materials preferentially use topography to arrest crack propagation. Indeed, one can compare man-made technologies for hard coatings (those in cutting tools, for instance) with a natural equivalent (teeth). For instance, the enamel/dentin interface combines compositional gradients with scalloped interfaces, which ensures stability. Corrugated interfaces are also observed in fish armour. Although there have been attempts to explore the effect of both topography and compositional and structural gradients on the mechanical properties of man-made materials, we have yet to match the structural complexity of natural materials.

At the microscopic level, natural composites are usually complex and anisotropic. They can have layered, columnar, or fibrous motifs. Quite often, the same structure can exhibit distinct layers with different motifs, such as the combination of columnar and lamellar regions in a shell. These motifs are usually orchestrated in sophisticated patterns, such as columns of circular layers in bone or wood, or the complex helicoidal arrangement of chitin fibres in the stomatopod club. Man-made composites can also be laminates or reinforced with complex fibre arrangements such as textile ceramic composites, but they have not yet attained the complexity of natural materials, which are characterized by features spanning many length scales.

Natural materials are often porous to provide paths for mass transport and/or to reduce weight. Furthermore, natural materials are usually graded or made of porous cores with dense shells to retain strength and flexibility. In some cases, such as bone osteons or dentin tubules, the pores play a significant role in toughening. Synthetic porous structures are usually crude in comparison; when high porosity is needed, it is usually at the expense of mechanical stability. The design of strong foams is now the subject of much investigation, and bioinspired hierarchical designs can offer efficient solutions.

Many natural materials are able to self-repair, often repeatedly and without external stimuli. In this regard, synthetic materials lag far behind. Although significant advances have been achieved in the area of self-healing structural composites, the number of repair cycles is often limited. To solve this problem, healing agents are delivered to the area of interest through vascular networks, or external stimuli (such as temperature) are used to trigger repair. There seems to be an inverse relationship between strength and the ability to repair autonomously multiple times.

5.2 Structure and Properties of Natural Materials

Biological materials are multifunctional. They combine biological, mechanical and other functions, and represent design solutions that are the local optimum for a given set of requirements and constraints. To separate mechanical from biological functions in natural materials, we derive material-property charts that represent sections through the multidimensional property space of materials and their performance [Fig.5.2(a)]. Such charts usually show specific properties—that is, normalized by density—because when size or weight are not relevant constraints, one can more readily attain both high strength and stiffness in a material.

Almost all natural materials are composites of some form, comprising a relatively small number of polymeric (proteins or polysaccharides, for example) and ceramic (for instance, calcium salts or silica) components or building blocks, which are often composites themselves. From this limited toolbox, an astonishing range of hybrid materials and structures are assembled. Wood, bamboo and palm, for example, comprise cellulose fibres within a lignin-hemicellulose matrix, shaped into hollow prismatic cells of varying wall thickness. Hair, nail, horn, wool, reptilian scales and hooves are formed from keratin, whereas insect cuticle consists of chitin in a protein matrix. The principal constituent of a mollusc shell is calcium carbonate, bonded with a few per cent of protein. Tooth enamel is composed of hydroxyapatite, while bone and antler are formed from hydroxyapatite and collagen. Collagen is the basic structural element for soft and hard tissues in animals, such as tendon, ligament, skin, fish scales, blood vessels, teeth, muscle and cartilage; in fact, the cornea of the eye is almost pure collagen.

When designing new materials, three factors are critical: chemical composition, nano/microstructure and architecture. Extensive manipulation of chemistry and microstructure is routinely required to make novel metallic alloys, ceramics, polymers and their composites. Throughout time, most advances in this area have occurred by trial-and-error experiments or through lucky accidents, as happened in the prehistoric Bronze Age and during its transition to the Iron Age. Conversely, evolutionary forces have led to the design concept of creating new materials with tailored properties through the manipulation of architecture, thereby permitting an enormous range of periodic, many-phase, continuous composites. For example, if we consider bone and nacre, both of which comprise rather meagre constituents in terms of their mechanical properties, the resulting natural composites display far superior properties [as exemplified by their fracture toughness in Fig.5.2(b)] that defeat the rule of mixtures. However, a pertinent question is whether we can emulate such designs and meet the greater challenge of making synthetic materials with such form and function.

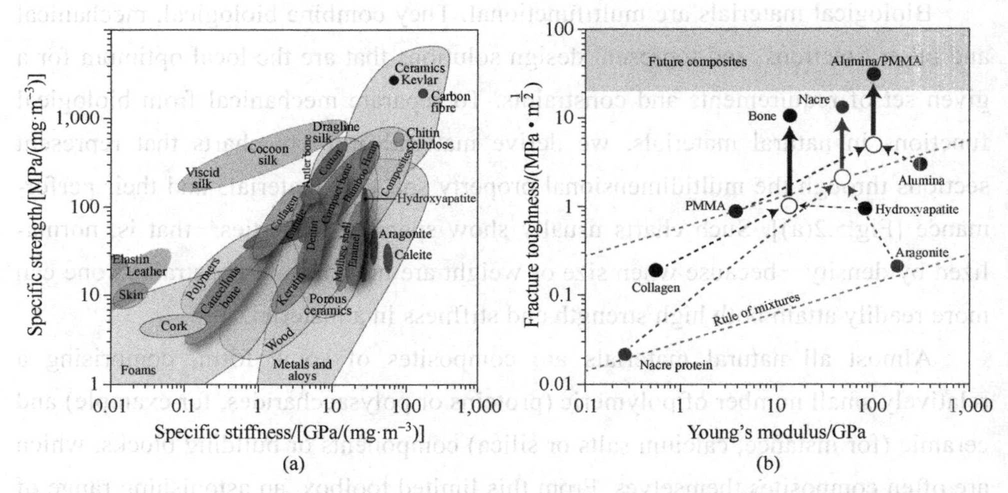

Fig.5.2 Material-property chart, and projections for natural and synthetic materials. (a) Ashby plot of the specific values (that is, normalized by density) of strength and stiffness (or Young's modulus) for both natural and synthetic materials. Many engineering materials, particularly high-performance ceramics and metallic alloys, have values of strength and toughness that are much higher than those of the best natural materials. Silk stands out as an exception, sometimes reaching the extraordinary toughness of 1,000 $MJ \cdot m^{-3}$ with a modulus of 10 GPa—approaching that of Kevlar. One might therefore conclude that there is nothing spectacular about the properties of natural materials, as in general they seem similar to what can be made synthetically. However, quite unlike most synthetic materials, all natural structural materials use a limited chemical palette of inexpensive ingredients—typically, proteins, polysaccharides, calcites and aragonites, and rarely metals—whose properties are often meagre, and are formed at ambient temperatures with little energy requirements. Moreover, the constituents of natural materials are typically arranged in a hierarchical architecture of interwoven or interlocking structures that is difficult to reproduce synthetically. Many natural materials also repair themselves when damaged; in contrast, self-healing synthetic structures are still highly limited. (b) Many natural composite materials, as exemplified by bone and nacre, have toughness values that far exceed those of their constituents and their homogeneous mixtures (as indicated by the dashed lines), and are able to sustain incipient cracking by utilizing extensive extrinsic toughening mechanisms. This results in much higher toughness for crack growth (closed symbols above the solid arrows) than for crack initiation (open symbols), and thus higher fracture toughness (solid arrows). By mimicking the architecture of nacre in a synthetic ceramic material (alumina/PMMA), similar behavior and exceptional toughness can be attained

One useful and frequently applied approach is that of materials design guided by first-principles calculations, which are based on an understanding of the different atomic components and the diverse sets of functional requirements. However, this strategy faces many issues. Despite decades of research, it has generally failed to yield significantly improved structural materials and performance, nor has it improved predictions of complex mechanical properties such as toughness. One reason for this failure is that current computational capabilities and tools are insufficient to be able to integrate into one model physical mechanism that acts at multiple length scales and that affects a material's mechanical performance. Moreover, some natural materials, such as bone, have the capacity for healing, self-repair and adaptation to changes in mechanical usage patterns, which pose an even greater challenge for the engineering of materials that mimic natural structures.

(1) Bone and nacre

As described in several recent review articles, bone and nacre (abalone shell) are prime examples of natural materials that combine strength with toughness, making them truly damage-tolerant. Also described in this subsection are other natural materials that have evolved efficient strategies for developing exceptional damage tolerance, including teeth, the dactyl clubs of stomatopod shrimps, and bamboo.

Bone is composed of cells embedded in an extracellular matrix, which is an ordered network assembled from two major nanophases: collagen fibrils made from type-I collagen molecules (~300 nm long, ~1.5 nm in diameter) and hydroxyapatite [$Ca_{10}(PO_4)_6(OH)_2$] nanocrystals (plate-shaped, 50 nm×25 nm in size, 1.5-4 nm thick) distributed along the collagen fibrils (Fig.5.3). The hydroxyapatite nanocrystals are preferentially oriented with their *c* axis parallel to the collagen fibrils, and arranged in a periodic, staggered array along the fibrils. These two nanophases make up about 95% of the dry weight of bone. The structures form a tough yet lightweight, adaptive, self-healing and multifunctional material. Bone derives its resistance to fracture with a multitude of deformation and toughening mechanisms operating at many size scales, ranging from the nanoscale structure of its protein molecules to the macroscopic physiological scale.

The origins of fracture resistance in healthy human cortical bone can be conveniently separated into intrinsic mechanisms that promote ductility and extrinsic mechanisms that act to "shield" a growing crack (Fig.5.1). Intrinsic toughening mechanisms result primarily from plasticity (Box 1). In bone, they originate from mechanisms at work at the smallest length scales, including molecular uncoiling of the mineralized collagen components and, most importantly, the process of fibrillar sliding. As load is applied, it is carried as tension in the mineral platelets and transferred between the platelets via shearing of the collagenous matrix. This fibrillar sliding mechanism is essential to promote plasticity at this length scale. Many aspects of the collagen fibril's structure, such as the hydroxyapatite/collagen interface, intermolecular crosslinking and sacrificial bonding, play a role in its ability to promote fibrillar sliding efficiently. These features constrain molecular stretching and provide the basis for the increased strength of the collagen. This makes a large regime of dissipative deformation possible once plastic yielding has begun. As in most materials, plasticity and the resultant ductility provide a major contribution to the intrinsic toughness by dissipating energy and forming plastic zones surrounding incipient cracks, which further serve to blunt crack tips, thereby reducing the driving force for cracking.

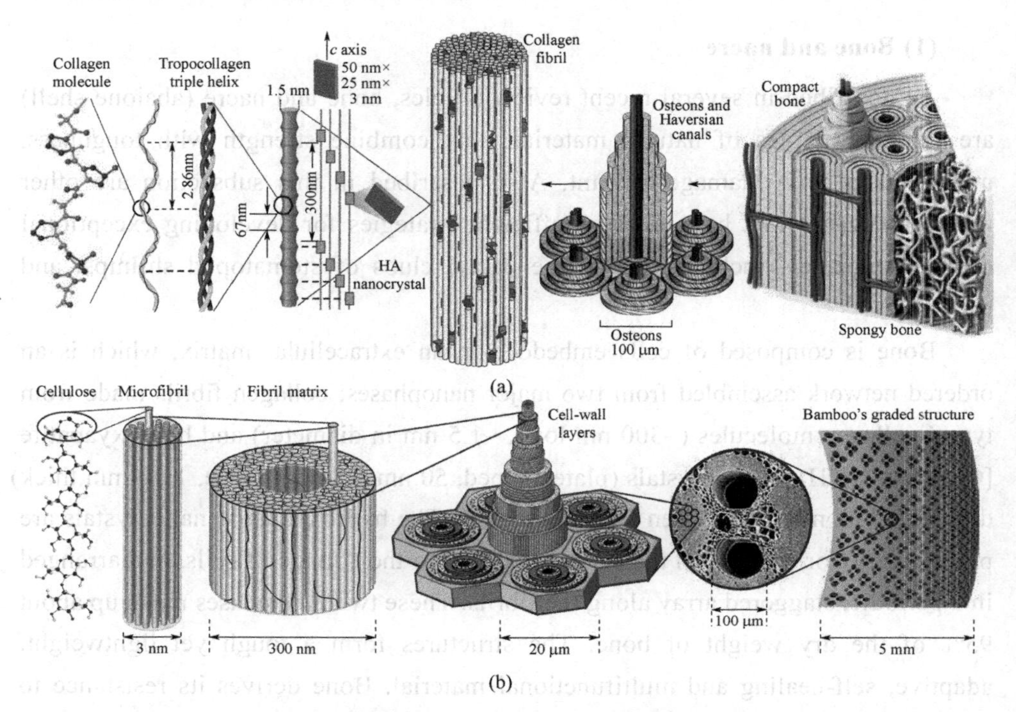

Fig.5.3 Hierarchical structure of bone and bamboo. (a) In bone, macroscale arrangements involve both compact/cortical bone at the surface and spongy/trabecular bone (foam-like material with ~100 μm-thick struts) in the interior. Compact bone is composed of osteons and Haversian canals, which surround blood vessels. Osteons have a lamellar structure, with individual lamella consisting of fibres arranged in geometrical patterns. The fibres comprise several mineralized collagen fibrils, composed of collagen protein molecules (tropocollagen) formed from three chains of amino acids and nanocrystals of hydroxyapatite (HA), and linked by an organic phase to form fibril arrays. (b) Bamboo is composed of cellulose fibres imbedded in a lignin-hemicellulose matrix shaped into hollow prismatic cells of varying wall thickness. In bamboo and palm, which have a more complex structure than wood, a radial density gradient of parallel fibres in a matrix of honeycomb-like cells increases each material's flexural rigidity. Bamboo increases its flexural rigidity even further by combining a radial density gradient with a hollow-tube cross-sectional shape

However, an even larger contribution to the fracture resistance of bone arises from mechanisms of extrinsic toughening at coarser length scales, in the range of ~10-100 μm. Specifically, once the crack begins to grow, mechanisms within the microstructure are activated to inhibit further cracking; indeed, the primary drivers for this are the nature of the crack path and its interaction with the bone-matrix structure. Two salient toughening mechanisms can be identified: crack bridging and crack deflection/twist. Crack bridging occurs as microcracks form ahead of the crack tip, primarily along the hypermineralized interfaces at the boundary of the osteons, and producing the so-called uncracked-ligament bridges, which act as intact regions spanning the crack wake to inhibit its progress. Crack deflection is particularly potent in the transverse orientation, where cracks are aligned perpendicular to the osteons. As the crack begins to grow, structural features such as osteocyte lacunae and porosity can deflect the crack path. However, the largest features, specifically secondary osteons and in particular their brittle interfaces (cement lines), are most effective at crack deflection. Such crack deflection toughens normal bone by diverting the crack path from the plane of maximum tensile stress; as such, crack-tip stress intensity decreases (typically by a factor of two or more), and a larger applied force is required to propagate the crack further. Indeed, it is because of this that the fracture toughness of bone, which in the longitudinal direction is typically 1-5 $\mathrm{MPa \cdot m^{\frac{1}{2}}}$ can be many times higher in the transverse direction,where cracks deflect at the cement lines. It is such extrinsic toughening, resulting in increased resistance to both initiated and growing cracks that is so effectively used in natural materials.

It should be noted that the small-scale intrinsic and larger-scale extrinsic processes are coupled. When the intrinsic toughness, generated at small length scales through fibrillar sliding, is degraded by biological factors (such as altered collagen crosslinking due to ageing and disease), the bone alternatively dissipates energy at higher length scales by microcracking. This is a form of plasticity that promotes extrinsic toughening at the microscale through the formation of deflected and bridged crack paths. Of course, a major characteristic of bone is its ability to remodel itself to repair damage—a trait that is difficult to replicate in synthetic materials. Indeed, there may be a coupling between bone inelasticity due to microcracking and the signalling that promotes such repair, as the microcracks are thought to serve the canaliculi, which are the means by which the osteocytes (osteoblasts that have become trapped within the bone matrix) remain in contact with other cells in bone.

Whereas bone is an example of extreme structural complexity, nacre exhibits similar properties with elegant simplicity. Essentially, both are laminated composites, containing hexagonal plates of either hydroxyapatite (bone) or aragonite. Nacre consists of 95 vol.% of layered aragonite ($CaCO_3$) platelets (~200-900 nm thick with a diameter of 5-8 μm), bonded by a thin (~10-50 nm thick) layer of organic material. However, its toughness is some three orders of magnitude higher than that of calcium carbonate. Actually, the fracture toughness of aragonite is ~0.25 $MPa \cdot m^{\frac{1}{2}}$, whereas the toughness of nacre does not exceed 10 $MPa \cdot m^{\frac{1}{2}}$, representing at best 40-fold increase in toughness (a factor of almost 2,000 in terms of energy).

The hard mineral aragonite provides for strength, but nacre would be brittle without a means of dissipating strain. Accordingly, inelastic deformation generated by limited interlayer shearing through the organic phase allows for such strain redistribution, thereby conferring toughness. The principal toughening mechanisms in nacre are crack bridging and the resulting "pull-out" of mineral bricks—associated with controlled, yet limited, sliding of the aragonite layers over each other—and aided by viscoplastic energy dissipation in the organic layer. There is still debate regarding the mechanisms that restrain sliding. Among those considered are resistance from the lamellae nanoroughness, plastic deformation of the aragonite at the nanometre level, the organic layer acting as a viscoelastic glue, the presence of (pre-existing) mineral bridges, and, for certain nacres, platelet interlocking at the microscopic level. As with bone, toughening in nacre is thus largely extrinsic and results in rising R-curve behaviour. The implication is that, unlike pure aragonite, a 95% aragonite composite material can tolerate the stable propagation of cracks.

Because many researches on bone and nacre have so far focused on structure-versus-property characterization as well as on how difficult it is to make synthetic materials in their image, it is easy to overlook that the making of these materials is the responsibility of cells. Indeed, the key to mimicking bone and nacre lies in understanding the involved cellular processes. However, many relevant biological issues remain unresolved. Both bone and nacre are soft-hard composites, and as such the roles of interfaces in the growth of these materials are crucial. Important steps in deciphering these roles have been the discovery of two new proteins that control the growth and crystal structure of aragonite, and how epithelial cells secrete all the necessary components for making the characteristic brick-and-mortar structure of nacre. Further progress in discerning the underlying mechanisms that control these biomineralization processes should yield clues for the design of novel biomimetic materials.

(2) Other natural structures (shrimp dactyl club, teeth, bamboo)

Reports describing the special characteristics of many natural materials now appear on a regular basis. However, success in truly mimicking the structures and properties of these materials is still extremely limited—particularly if they must be processed in bulk form. Nevertheless, there has been significant progress in discerning some of the unique mechanisms. A case in point is a recent study on harlequin (peacock) mantis shrimp, which claimed that the dactyl club (biological hammer) of this creature has a much higher specific strength and toughness than those of any synthetic composite material. In fact, the force generated by the club can reach 500 N, which is strong enough to break aquarium glass. These unusual properties and outstanding impact resistance are achieved by an elastic modulus mismatch that is controlled by the amount of mineral phase (crystalline and amorphous hydroxyapatite). The structure of the dactyl club consists of three distinct regions within an organic matrix, all comprising multiphase composites of oriented crystalline hydroxyapatite and amorphous calcium phosphate and carbonate. Each region has comparable thickness (~75 μm), separated by thin layers of protein (chitin). The resulting graded, layered structure, wherein each region has a different elastic modulus, forces any incipient cracks to continually change direction, thereby retarding crack growth via the extrinsic toughening mechanism of crack deflection.

5.3 Methods of Processing Hierarchical Materials

The large span in length scales and the overall complexity of designing new biomimetic materials impose a combination of requirements and design motifs (Box 2) that are well beyond the reach of present technology. On the one hand, the basic knowledge required to meet these requirements is incomplete in many dimensions, in particular at the nanoscale. On the other hand, the need for novel processing routes to address multiple physical and mechanical factors imposes tremendous constraints on the choice of suitable materials. The ultimate objective is to develop smart materials that can both detect an event and structurally change in response to it. We are far from being able to achieve this goal today, but a concerted and well-directed effort may get us there within 10-20 years.

(1) Biomimetic mineralization

All hard materials in animals (including humans) are mineralized from a limited number of inorganic components—calcium carbonate, silica (glass) and calcium phosphate being the most common. To build hard materials, living organisms take advantage of biomineralization, a process in which dedicated cells deposit minerals to a soft polymeric (protein) matrix to strengthen, harden and/or stiffen it. This shows that biomimetic mineralization could be, in principle, an obvious and effective process for building synthetic materials. However, the results reported so far are disappointing because the techniques are slow and only able to produce micrometre-sized specimens with mechanical properties that fall short of those of their natural counterparts. In spite of this, the mineralization process has been used to grow particles, capsules and films with control of mineral morphology, size and even the crystalline phase. The mineralization of an organic scaffold to create a practical three-dimensional (3D) material is particularly challenging. It requires a combination of the correct scaffold chemistry (to template mineral formation) with a suitable process that reproduces the role of cells. The former is usually achieved by creating organic scaffolds with ion-binding sites that promote heterogeneous nucleation. Numerous strategies have been used, from the use of natural polymers—some of which are known to regulate mineralization in organisms—to the application of phage display technologies to screen for optimum peptides that promote mineral nucleation, to the specific design of synthetic organic matrices. Mineralization on carbon nanostructures, such as graphene or nanotubes, has also been investigated. Mineral nucleation requires ion supersaturation, which can be reached, for example, through a controlled and homogeneous change in the pH of the solution in which the scaffold is immersed. However, cells often promote localized mineralization using processes, such as the secretion of vesicles, which are difficult to replicate in the laboratory. Furthermore, although much work has focused on the control of mineral growth and structure at the nanoscale, manufacturing techniques that use biomimetic mineralization for fabricating complex composites with features ranging from the micro to the macroscale (as required for structural applications) are still far from being practical.

An intriguing approach for the formation of complex structures with several organization hierarchies is the combination of controlled mineral growth and subsequent particle self-assembly following external stimuli. In this approach, smart responsive particles are prepared through functionalization by using either inorganic nanoparticles or organic molecules. External stimuli can then be used to control the assembly of the functionalized particles or their distribution within a composite. For instance, it has been demonstrated that how functionalization with organic molecules can be used to promote recognition and self-assembly so as to build ordered structures. Janus particles—particles that display different chemistry on opposite sides of their surface—can be used to obtain complex hierarchical arrangements. Progressing from particle assembly at the nanoscale and microscale to making a practical structural material still represents a major problem, although there have been some significant advances. For example, magnetic fields can be used to control the distribution of a magnetic reinforcement in a polymer. By using magnetic fields on ceramic platelets functionalized with magnetic nanoparticles at their surface, composites with complex arrangements of these platelets within an organic matrix have not only been made with practical dimensions, but also manipulated to achieve a desired mechanical response. It has also been possible to create macroscopic structures using particles that can assemble in emulsions. In some cases, this assembly was made reversible by functionalizing the particles with pH-responsive polymers. However, the extent to which it is possible to mimic natural structures using these techniques is still a matter of debate.

(2) Freeze casting

A more recent attempt at mimicking nacre in a bulk material involved the use of freeze casting (also known as ice templating) to provide a new class of bioinspired ceramic materials with exceedingly high toughness. This technique is a relatively inexpensive procedure for the controlled freezing of ceramic-based suspensions in water, and as such provides a means to mimic natural structural designs over several length scales. Specifically, ceramic suspensions are directionally frozen under conditions designed to promote the formation of lamellar ice crystals, which expel the ceramic particles as they grow. After sublimation of the water, this results in a layered homogeneous ceramic scaffold that, architecturally, is a negative replica of the ice. The scaffold can then be filled with a second soft phase so as to create a hard-soft layered composite. It is also possible to create brick-and-mortar structures by compressing the layered ceramic scaffolds before the introduction of the soft phase. Also, by controlling the freezing kinetics and the composition of the suspension, the architecture of the material can be adjusted at several length scales so as to replicate some of the microstructural features responsible for the unique mechanical response of like-nacre materials. Furthermore, the thickness of the lamellae can be manipulated by controlling the speed of the freezing front, such that materials with lamellae as thin as 1 μm (close to that of nacre) ordered over macroscopic dimensions can be fabricated in practical sizes. The roughness of the lamellae (which replicates that of the ice crystals) can be engineered at submicrometre to micrometre levels by using additives such as sucrose, salts or ethanol, which vary the solid-liquid interfacial tension and the phase diagram of the solvent. Moreover, the growing ice crystals can split and trap ceramic particles, which generate inorganic bridges between lamellae. All of these factors, together with the roughness of the ceramic, the properties of the soft phase and the strength of the hard/soft interface, control how much the ceramic bricks can slide relative to each other, which, as with natural nacre, is the dominant mechanism controlling the ductility of the material.

Freeze casting has been used to fabricate metal/ceramic and model polymer/ceramic [poly (methyl methacrylate), PMMA]/alumina and, more recently, PMMA/SiC hybrid materials with fine lamellar or brick-and-mortar architectures—the latter with high ceramic contents up to 80 vol.%. In these cases, the objective was to allow for stress relaxation by combining a strong and hard (ceramic) phase with a compliant (polymer or metal) layer in between. The nacre-like PMMA/alumina materials in particular showed remarkable R-curve behavior, thus indicating that they are tolerant to the stable growth of cracks, with a fracture toughness of 30 $MPa \cdot m^{\frac{1}{2}}$ or more and yield strengths of 200 MPa (some 300 times higher in energy terms than those of either the constituent ceramic or polymer). Furthermore, they were shown to be 50%-100% more ductile than nacre, and twice as strong. In fact, such hybrid alumina ceramics approach the specific strength and toughness of aluminum metal alloys while exhibiting lower density and higher stiffness. Strong and tough ceramics have also been fabricated through the freeze casting of platelet-containing suspensions, which opens new questions regarding the role of the soft phase in nacre-like materials.

Freeze casting can also be used for making efficient cellular structures. Although foaming techniques are well established, freeze casting and 3D printing both offer many advantages in creating such structures. For example, directional solidification during freeze casting of platelet-based slurries can be used for the development of highly porous honeycomb-like scaffolds with a nacre-like cell-wall structure, which arises from the self-assembly of the ceramic platelets during processing.

(3) Additive manufacturing

This type of manufacturing encompasses a family of technologies that draw on computer designs to build structures layer by layer. Different approaches have been developed, such as droplet delivery (typically, 3D inkjet printing), continuous extrusion (for example, robocasting), selective laser sintering, and the use of light to cure designed areas in polymer-containing suspensions (for instance, stereolithography or two-photon polymerization). Additive-manufacturing technologies have been used to build functional networks of cellular materials and complex devices such as batteries, photonic crystals, tissue-engineering scaffolds, catalyst supports and foams for acoustic-, vibration- or shock-energy-damping applications.

In some respects, it may be ideal to use additive manufacturing to assemble, on demand, structures modelled after natural materials. However, the practical fabrication of bioinspired composites with these techniques will hinge on the ability to solve some difficult problems. First, the palette of materials that can be processed by additive manufacturing ought to be broadened. Only a number of metals, polymers and in particular a limited number of ceramics can now be used to build structures with features ranging in size from tens of micrometers to one submicrometer, depending on the technology. In fact, the high thermal stability of ceramics in part hampers the use of techniques that involve melting or in situ sintering. Most ceramic additive-manufacturing technologies require an "ink", typically a colloidal suspension in water or other solvent, or a wax containing ceramic particles. Furthermore, the parts usually require additional thermal treatments for consolidation. To make things more complicated, bioinspired materials are usually hybrids that combine dissimilar materials (for example, a polymer and a ceramic)—something that may be difficult to construct using a single technique.

Second, the precision required to print nanoscale features ought to be combined with the fabrication of large-scale components. On the one hand, continuous extrusion of sol-gel ceramic inks, two-photon polymerization (nanolithography) or inkjet printing can be used to build materials with fine features; however, they cannot be used to create large structures. On the other hand, techniques such as robocasting using colloidal ceramic inks, 3D printing or stereolithography, offer the potential for large-scale manufacturing, yet their ultimate feature resolution is of the order of tens or hundreds of micrometers.

Despite these difficulties, there has been some progress. For example, recent studies have demonstrated that tissue constructs that mimic living tissue can be made by printing tens of thousands of picolitre droplets in 3D. Also, additive manufacturing has been used to fabricate model structures inspired by natural composites, such as nacre or the stomatopod club. Although the characteristic features of such model composites are usually orders of magnitude larger than in their natural counterparts, they provide a powerful platform for isolating and testing a specific design concept taken from nature.

Additive manufacturing may provide a path towards fabricating the bioinspired structural materials of the future. Much work is still needed, particularly because the current "Achilles heel" of such techniques is the difficulty of controlling the surface quality and microstructure of individual layers and segments, thus hampering the reliable combination of mechanical properties required for the structural application of the finished materials.

5.4 Looking Ahead

After decades of research, it seems that many structural materials are quickly approaching their performance limits. For example, Ni-based superalloys in jet-engine gas turbines have reached their temperature ceiling. Additionally, there is an unmet demand for lighter, tougher and more wear-resistant materials across a range of applications, from minimally invasive orthopaedic implants to high-efficiency industrial cutting and drilling tools. Where do we go from here? What will inspire new ideas? Bioinspired design clearly provides one feasible route. In the natural world, multiple-scale, multifunctional and hierarchically organized structures result in hybrid materials with unusual, often remarkable, properties acquired through combinations of relatively mundane constituents. Although we can draw on natural materials themselves for use in applications at ambient conditions, the challenge is to translate insights gained from observing their complex architecture into structural materials that can operate under extreme environments, in ever increasing temperatures and pressures. Another noteworthy observation is that natural structures are usually optimized for specific applications. The microstructure of enamel, for instance, varies from animal to animal or even from tooth to tooth. However, synthetic materials usually see a wider range of applications. Aerospace alloys, for example, are also used in orthopaedic implants. Will bioinspired synthetic materials also become highly specialized? If so, economic considerations are likely to become as important as performance.

In basic terms, there are two fundamental challenges facing researchers in the field of bioinspired materials: ①to devise approaches for translating design motifs found in natural materials to a wider range of material combinations; ②to create methods for making bioinspired structural materials in practical form and in bulk.

The first of these challenges is exceedingly complicated. For example, can the complex role of nacre's protein film be captured by a thin metal or carbon layer? Whereas the soft components in materials such as nacre or silk are built of molecules designed to unravel in a controlled manner so as to accommodate significant deformations, the thin, confined metallic layers may exhibit fairly limited plastic deformation. It has thus been suggested that nanoporous ceramic interlayers can be used to replicate the role of the protein layer in nacre, and recent work shows how high fracture resistance can be achieved in glass by guiding the crack path through interlocking interfaces without the addition of soft phases. The translation of natural design motifs will therefore require a deep understanding of the mechanical behaviour of materials and interfaces in volumes down to the nanoscale. This is an area of research that has attracted much interest in recent years, but we are still far from understanding how natural materials are built or develop their properties. In this respect, the inherent hierarchical nature of biological materials demands new multiscale modelling approaches to capture fully the relationship between structure and performance. This is a relatively new field that is already yielding much insight into, for example, the parameters that control the behaviour of brick-and-mortar structures. Indeed, the nature of the adhesion at the hard/soft interface, as well as the structure of the bricks and the properties of the mortar, has been recently explored.

The second challenge is primarily focused on manufacturing. New processing techniques, such as those reviewed here, undoubtedly offer much early promise, yet more effort must be devoted to them before their potential can be realized. Integrated computational modelling of both the manufacturing process and resultant properties will be important, despite the fact that the ability of these techniques to predict critical mechanical properties, in particular fracture toughness, fatigue or wear resistance, is still a long way off. The key feature responsible for the success of natural materials is the precise way in which their components are arranged in space (architecture) across numerous length scales. The natural world provides us with a blueprint to emulate, yet we must develop our own methods of construction. The synthesis of nacre-like or bone-like materials is an attractive goal, but we have yet to demonstrate a system that is sufficiently durable or can be manufactured at a reasonable cost. Furthermore, it is possible that the term "bioinspired" is being abused. Many layered materials available to engineers today are quite far from the structure of nacre, and not all porous foams imitate the architecture of bone or wood. The development of practical manufacturing approaches is further off than many researchers would like to believe.

A common theme when reviewing current processing technologies is the lack of systematic methodology for dealing with the structural complexity of a truly bioinspired structure whose dimensions span from the nanoscale to the macroscale. Traditional materials science has focused on the microscale; more recently, nanotechnology has opened new opportunities. However, it remains a dauntingly complex problem to develop a manufacturing technique that will allow flexible control of nanometric features in structures built at large scales. One solution is to combine two or more fabrication technologies. For example, can 3D-printing techniques be used to build scaffolds that can be subsequently mineralized? An example of this is the combination of solid free-form fabrication with thin-film deposition techniques to create ultralightweight metallic and ceramic microlattices; in fact, initial studies have demonstrated that significant improvements in mechanical efficiency can be achieved. An alternative solution is to apply a reductionist strategy: to fabricate large-scale structural components through the controlled assembly of small building blocks or modules. These can be defined as homogeneous and viable constructs, built with nanometre-level precision using existing techniques, and with surfaces that are specifically engineered to interconnect in a tightly controlled manner and provide specific functions. Using such modules, complex hierarchical structures can be built from simple units, and complex processes can be orchestrated that mimic strategies by which natural materials are assembled from smaller modules, each designed to perform specific functions.

Beyond their passive mechanical response, natural structures are active in that they respond and react autonomously to external conditions, and are able to self-repair, often repeatedly. Although these possibilities are outside the scope of this review, scientists and engineers are already researching ways to replicate these traits in engineered composites, from the use of microvascular networks to deliver healing agents, to the design of self-shaping composites.

Natural structures are wonderful examples of what can be done with a fairly limited selection of materials. Humans have enriched the natural-materials palette with thousands of synthetic compounds. Perhaps because there is such an abundance of synthetic materials, we have not been compelled to refine their designed architectures with the same degree of sophistication. Compared with the complexity of natural materials, many synthetic materials may in fact seem crude. With the advantage of time, natural evolution has led to nano- and micro-structures as advanced and sophisticated as the large-scale designs of bridges or buildings. Bioinspiration thus poses a simple and enticing challenge: how far can we go in manipulating structure? This is not just a test of human ingenuity; the potential pay-out is huge. The possibility of marrying the structural control found in nature with the huge variety of synthetic compounds could lead to the development of new materials, extend the range of application of current ones, and break existing limitations in terms of weight, toughness, strength and environmental resistance. Attaining those goals would be a major technological advance. In the end, regardless of whether bioinspired materials are widely implemented, even those skeptics of their promise must recognize that understanding natural structures and finding ways to replicate them will yield insights that can drive materials science forward for years to come.

Key vocabularies

abalone [ˌæbəˈləʊni] ***n.*** 鲍鱼

Achilles heel [əˌkɪliːz ˈhiːl] ***n.*** 薄弱环节，要害

acoustic [əˈkuːstɪk] ***adj.*** 声音的，听觉的

additive manufacturing ***n.*** 增材制造；添加剂制造

amorphous [əˈmɔːfəs] ***adj.*** 无组织的；模糊的；无固定形状的；非结晶的
antler [ˈæntlə(r)] ***n.*** 鹿角；角枝
aquarium [əˈkweəriəm] ***n.*** 水族馆；水族槽，养鱼缸；玻璃鱼池
aragonite [əˈrægənaɪt] ***n.*** 霰石，文石
architecture [ˈɑːkɪtektʃə(r)] ***n.*** 建筑设计，建筑学；结构，架构
arrest [əˈrest] ***v.*** 阻止，抑制；吸引；心跳停止．***n.*** 停止，终止
beak [biːk] ***n.*** 鸟喙；鹰钩鼻；掌权者
calcite [ˈkælsaɪt] ***n.*** 方解石
canal [kəˈnæl] ***n.*** 运河，灌溉渠，水道；（体内的）管，道．***v.*** 在……开凿运河
canaliculi [ˌkænəˈlikjulai] ***n.*** （骨头或植物某些部位的）微管（canaliculus 的复数）
cancellous [ˈkænsələs] ***adj.*** 多孔的；罗眼状的（等同于 cancellate）
cartilage [ˈkɑːtɪlɪdʒ] ***n.*** 软骨；软骨结构
catastrophic [ˌkætəˈstrɒfɪk] ***adj.*** 灾难性的；极糟的，失败的；大规模突变的
cellulose [ˈseljuləʊs] ***n.***（植物的）细胞膜质，纤维素，纤维素化合物
cement [sɪˈment] ***n.*** 水泥；胶合剂，接合剂．***v.*** 黏结，胶合；巩固，加强；决定，确立
ceramic [səˈræmɪk] ***n.*** 陶瓷制品，陶瓷器；制陶艺术．***adj.*** 陶瓷的
chitin [ˈkaɪtɪn] ***n.*** 壳质，几丁质，角素，甲壳素
collagen [ˈkɒlədʒən] ***n.*** 胶原蛋白，胶原质
colloidal [kəˈlɔɪdl] ***adj.*** 胶体的；胶质的；胶状的
concert [ˈkɒnsət] ***v.*** 共同议定，协调
confer [kənˈfɜː(r)] ***v.*** 授予，赋予；商讨，交换意见
confluence [ˈkɒnfluəns] ***n.*** 汇流点；聚集，合并
consolidation [kənˌsɒlɪˈdeɪʃən] ***n.*** 巩固；合并
constitutive law ***n.*** 本构关系，本构律，组成律
contemplate [ˈkɒntəmpleɪt] ***v.*** 沉思，盘算，打算；注视；考虑接受
conviction [kənˈvɪkʃn] ***n.*** 定罪，判罪；坚定的信仰，深信的观点；确信，深信
cornea [ˈkɔːniə] ***n.*** 眼角膜
corrugate [ˈkɒrʊgeɪt] ***vt.*** 使起皱；成波状．***adj.*** 起皱的；波状的．***vi.*** 起皱；缩成皱状
cortical [ˈkɔːtɪkəl] ***adj.*** 皮层的，皮质的，有关脑皮层的
counterpart [ˈkaʊntəpɑːt] ***n.*** 职能（或地位）相当的人；对应的事物
cuticle [ˈkjuːtɪkl] ***n.*** 外皮；角质层；表皮；护膜
dactyl [ˈdæktɪl] ***n.*** 趾
damping [ˈdæmpɪŋ] ***n.*** 阻尼；减幅；衰减
decipher [dɪˈsaɪfə(r)] ***v.*** 破译，辨认；理解
deflection [dɪˈflekʃn] ***n.*** 偏斜，偏移，歪曲；偏斜度，偏角
dentin [ˈdentɪn] ***n.*** 象牙质；牙质（等同于 dentine）

discern [dɪˈsɜːn] ***v.*** 看出，觉察出；了解，认识
displacement [dɪsˈpleɪsmənt] ***n.*** 取代，代替；排水量；电位移
ductility [dʌkˈtɪlɪtɪ] ***n.*** 展延性，柔软性；顺从；韧性；塑性；展性
duplicate [ˈdjuːplɪkeɪt,ˈdjuːplɪkət] ***v.*** 复制，复印；重复
embrace [ɪmˈbreɪs] ***v.*** 拥抱；欣然接受，乐意采纳；包括，涉及；围绕，环绕
emulate [ˈemjuleɪt] ***v.*** 仿效，模仿；仿真；努力赶上
enamel [ɪˈnæml] ***n.*** 搪瓷；珐琅；瓷釉；釉质；指甲油．***vt.*** 彩饰；涂以瓷釉
encompass [ɪnˈkʌmpəs] ***vt.*** 围绕，包围；包含或包括某事物；完成
endow [ɪnˈdaʊ] ***v.*** 赋予
entic [ɪnˈtaɪs] ***vt.*** 诱惑；怂恿
epithelial [ˌepɪˈθiːlɪəl] ***adj.*** 上皮的；皮膜的
exemplify [ɪɡˈzemplɪfaɪ] ***v.*** 是……的典范；举例说明
extrinsic [eksˈtrɪnsɪk] ***adj.*** 非本质的；外在的；外来的；外部的
fibril [ˈfaɪbrɪl] ***n.*** 纤丝，原纤维；须根
fracture toughness ***n.*** 断裂韧性
harlequin [ˈhɑːləkwɪn] ***n.*** 花斑眼镜蛇
heterogeneous [ˌhetərəˈdʒiːniəs] ***adj.*** 由很多种类组成的；不均一的，多相的
hexagonal [heksˈæɡənl] ***adj.*** 六角形的，六边形的
hierarchical [ˌhaɪəˈrɑːkɪkl] ***adj.*** 按等级划分的，等级（制度）的；分层的
homogeneous [ˌhɒməˈdʒiːniəs] ***adj.*** 同种类的，同性质的；齐性的，齐次的
hydroxyapatite [haɪdrɒksɪˈæpətaɪt] ***n.*** 羟磷灰石
incipient [ɪnˈsɪpiənt] ***adj.*** 初期的；初始的；起初的；发端的
intact [ɪnˈtækt] ***adj***. 完整无缺的，未经触动的，未受损伤的；原封不动的
interplay [ˈɪntəpleɪ] ***n.*** 相互影响，相互作用．***vi.*** 相互影响，相互作用
intricate [ˈɪntrɪkət] ***adj.*** 错综复杂的；难理解的
intrigue [ɪnˈtriːɡ] ***n.*** 密谋，阴谋；私通，奸情．***vi.*** 耍阴谋．***vt.*** 以谋略达成；激起……的好奇心
invasive [ɪnˈveɪsɪv] ***adj.*** 扩散性的，侵入的；切入的，开刀的
keratin [ˈkerətɪn] ***n.*** 角蛋白，角素，角质
lacunae [ləˈkjuːniː] ***n.*** 空白（名词 lacuna 的复数形式）；空隙，腔隙；缺陷
lamella [ləˈmelə] ***n.*** 薄层；薄片；瓣；鳃
lamellar [ləˈmelə] ***adj.*** 薄片状的，薄层状的
laminate [ˈlæmɪnət] ***v.*** 锻压成薄片；分成薄片．***adj***. 薄片状的；层状的
ligament [ˈlɪɡəmənt] ***n.*** 韧带；纽带
lignin-hemicellulose ***n.*** 木质素-半纤维素
mantis [ˈmæntɪs] ***n.*** 螳螂
meagre [ˈmiːɡə(r)] ***adj.*** 瘦的；贫弱的；思想贫乏的；土地不毛的
mesoscale [ˈmesəskeɪl] ***n.*** 中尺度；介观尺度．***adj.*** 中尺度的

modulus [ˈmɒdjʊləs] ***n.*** 系数，模数
mollusc [ˈmɒləsk] ***n.*** 软体动物
mundane [mʌnˈdeɪn] ***adj.*** 单调的，平凡的，平淡的；世俗的，尘世的
nacre [ˈneɪkə] ***n.*** 珍珠层；珠母贝；珍珠母
onset [ˈɒnset] ***n.*** 开始，发作，初现
orchestrate [ˈɔːkɪstreɪt] ***vt.*** 把（乐曲）编成管弦乐；和谐地安排；精心策划
orthopaedic [ˌɔːθəˈpiːdɪk] ***adj.*** 整形外科的，矫形的
osteoblast [ˈɒstɪəblæst] ***n.*** 成骨细胞；造骨细胞
osteocyte [ˈɒstɪəsaɪt] ***n.*** 骨细胞
osteon [ˈɒstɪɒn] ***n.*** 骨单位
palette [ˈpælət] ***n.*** 调色板；一组颜色
peptide [ˈpeptaɪd] ***n.*** 肽，缩氨酸
pertain [pəˈteɪn] ***v.*** 适合，关于，适用
phosphate [ˈfɒsfeɪt] ***n.*** 磷酸盐
picolitre ***n.*** 兆分之一升，皮升
platelet [ˈpleɪtlət] ***n.*** 血小板；小盘；小板
polysaccharide [ˌpɒlɪˈsækəraɪd] ***n.*** 多糖，多聚糖
potent [ˈpəʊtnt] ***adj.*** 强大的，有力的
prismatic [prɪzˈmætɪk] ***adj.*** 棱柱的，棱镜的；五光十色的，光彩夺目的
reptilian [repˈtɪlɪən] ***adj.*** 爬行动物的；爬虫的．***n.*** 两栖动物
salient [ˈseɪlɪənt] ***adj.*** 显著的，突出的；跳跃的；（指角）凸出的
scallop [ˈskɒləp] ***n.*** 扇贝，干贝．***vt.*** 使成扇形．***vi.*** 拾扇贝
silica [ˈsɪlɪkə] ***n.*** 硅石，二氧化硅；硅氧
sinter ['sɪntə] ***n.*** 烧结物；熔渣．***vt.*** 使烧结；使熔结．***vi.*** 烧结；熔结
sol-gel ***n.*** 溶胶-凝胶
specific strength ***n.*** 比强度；强度系数
stagger [ˈstægə(r)] ***adj.*** 交错的，错开的
stereolithography ***n.*** 立体平版印刷
stomatopod ['stəʊmətəpɒd] ***n.*** 口脚类动物；口足
strut [strʌt] ***n.*** 支杆；支柱
sublimation [ˌsʌblɪˈmeɪʃn] ***n.*** 升华，升华作用；升华物
sucrose [ˈsuːkrəʊz] ***n.*** 蔗糖
tendon [ˈtendən] ***n.*** 筋，腱
topography [təˈpɒgrəfi] ***n.*** 地形学，地形测量学；地形，地貌；局部解剖（图）
toughness [ˈtʌfnəs] ***n.*** 韧性；硬，切不动；坚韧，坚强；艰巨性；强硬
trabecular [trəˈbekjʊlə] ***adj.*** 小梁的；有小带的；有横隔片的
tropocollagen [ˌtrəʊpəʊˈkɒlədʒɪn] ***n.*** 原胶原；原胶原蛋白
tubule [ˈtjuːbjuːl] ***n.*** 小管，细管；菌管

two-photon polymerization ***n.*** 双光子聚合技术
uniaxial [ˌjuːnɪˈæksɪəl] ***adj.*** 单轴的；单轴晶体
unravel [ʌnˈrævl] ***vt.*** 解开；阐明；拆散. ***vi.*** 被解开；被拆散
vesicle [ˈvesɪkl] ***n.*** 囊，泡，小泡
viscoelasticity [ˈvɪskəʊɪlæsˈtɪsɪtɪ] ***n.*** 黏弹性
yet-to-be-developed ***adj.*** 尚未开发的
zirconia [zəˈkəʊnɪə] ***n.*** 氧化锆

Exercises

Ⅰ. Translate the following English into Chinese.

1. Our examination of this issue begins with a brief review of important natural structural materials and the mechanisms underlying their mechanical behavior and function, and is followed by a detailed discussion of the key lessons offered by these materials and of the difficulties encountered in attempts to implement them in practical synthetic structures.

2. Numerous strategies have been used, from the use of natural polymers—some of which are known to regulate mineralization in organisms—to the application of phage display technologies to screen for optimum peptides that promote mineral nucleation, to the specific design of synthetic organic matrices.

3. All of these factors, together with the roughness of the ceramic, the properties of the soft phase and the strength of the hard/soft interface, control how much the ceramic bricks can slide relative to each other, which, as with natural nacre, is the dominant mechanism controlling the ductility of the material.

4. Much work is still needed, particularly because the current “Achilles heel” of such techniques is the difficulty of controlling the surface quality and microstructure of individual layers and segments, thus hampering the reliable combination of mechanical properties required for the structural application of the finished materials.

5. This is not just a test of human ingenuity; the potential pay-out is huge. The possibility of marrying the structural control found in nature with the huge variety of synthetic compounds could lead to the development of new materials, extend the range of application of current ones, and break existing limitations in terms of weight, toughness, strength and environmental resistance.

Ⅱ. Answer the following questions based on the main text in this chapter.

1. What is the constitutive law of material?

2. How do natural materials well integrate strength and toughness?

3. What are the critical factors when developing new materials?

4. What is the underlying mechanism of Mantis shrimp's dactyl club that has so much higher specific strength and toughness than any synthetic composite material?

5. Can you talk about the advantages of the additive manufacturing method when fabricating advanced functional materials?

Ⅲ. Extensive reading.

Yayun Wang，Steven E Naleway，Bin Wang. Biological and bioinspired materials: Structure leading to functional and mechanical performance[J]. Bioactive Materials, 2020, 5(4), 745-757.

Chapter 6

Bioinspired Robots

Human inventors and engineers have always found in nature's products an inexhaustible source of inspiration. About 2400 years ago, for instance, Archytas of Tarentum allegedly built a kind of flying machine, a wooden pigeon balanced by a weight suspended from a pulley, and set in motion by compressed air escaping from a valve. Likewise, circa 105 AD, the Chinese Cai Lun is credited with inventing paper, after watching a wasp create its nest. More recently, Antoni Gaudi's design of the still-unfinished Sagrada Familia cathedral in Barcelona displays countless borrowings from mineral and vegetal exuberance.

Although a similar tendency underlied all attempts at building automata or protorobots up to the middle of the last century, in the last decades, roboticists borrowed much more from mathematics, mechanics, electronics, and computer science than from biology. On the one hand, this approach undoubtedly solidified the technical foundations of the discipline and led to the production of highly successful products, especially in the field of industrial robotics. On the other hand, it served to better appreciate the gap that still separates a robot from an animal, at least when qualities of autonomy and adaptation are sought. As such qualities are required in a continually growing application field—from planetary exploration to domestic uses, a spectacular reversal of interest towards living creatures can be noticed in current-day robotics, up to the point that it has been said that natural inspiration is the "new wave" of robotics.

Undoubtedly, this new wave would not have been possible without the synergies generated by recent advances in biology where so-called integrative approaches now produce a huge amount of data and models directly exploitable by roboticists, and in technology with the massive availability of low-cost and power-efficient computing systems, and with the development of new materials exhibiting new properties. This will be demonstrated in this article that: first, reviews recent research efforts in bioinspired morphologies, sensors, and actuators;

then, control architecture that, beyond mere reflexes, implement cognitive abilities like memory or planning or adaptive processes—like learning, evolution and development will be described; finally, the article will also report related works on energetic autonomy, collective robotics, and biohybrid robots.

It should be noted that this chapter will describe both bioinspired and biomimetic realizations. In fact, these two terms respectively characterize the extremities of a continuum for which, on the one side, engineers seek to reproduce some natural result, but not necessarily the underlying means, while, on the other side, they seek to reproduce both the results and the means. Thus, bioinspired robotics tends to adapt to traditional engineering approaches, some principles that are abstracted from the observation of some living creature, whereas biomimetic robotics tends to replace classical engineering solutions with a detailed mechanisms or processes that it is possible to reproduce from the observation of this creature. In practice, any specific application usually lies somewhere between these two extremities. Be that as it may, because biomimetic realizations are always bioinspired, whereas the reverse is not necessarily true, qualifying expressions like bioinspired or biologically-inspired will be preferentially used in this chapter.

6.1 Bioinspired Morphologies

Although not comparable to that of real creatures, the diversity of bioinspired morphologies that may be found in the realm of robotics is nevertheless quite impressive. Actually, a huge number of robots populate terrestrial, as well as aquatic or aerial environments, and look like animals as diverse as dogs, kangaroos, sharks, dragonflies, or jellyfishes, not to mention humans (Fig.6.1).

In nature, the morphology of an animal fits its ecology and behavior. In robotics applications, bioinspired morphologies are seldom imposed by functional considerations. Rather, as close a resemblance as possible to a given animal is usually sought per se, as in animatronics applications for the entertainment industry. However, several other applications are motivated by the functional objective of facilitating human-robot interactions, thus allowing, for instance, children or elderly people to adopt artificial pets and enjoy their company. Such interactions are facilitated in the case of so-called anthropopathic or human-friendly robots, like Kismet at MIT or WE-4RII at Waseda University, which is able to perceive and respond to human emotions, and express apparent emotions influencing their actions and behavior (Fig.6.2).

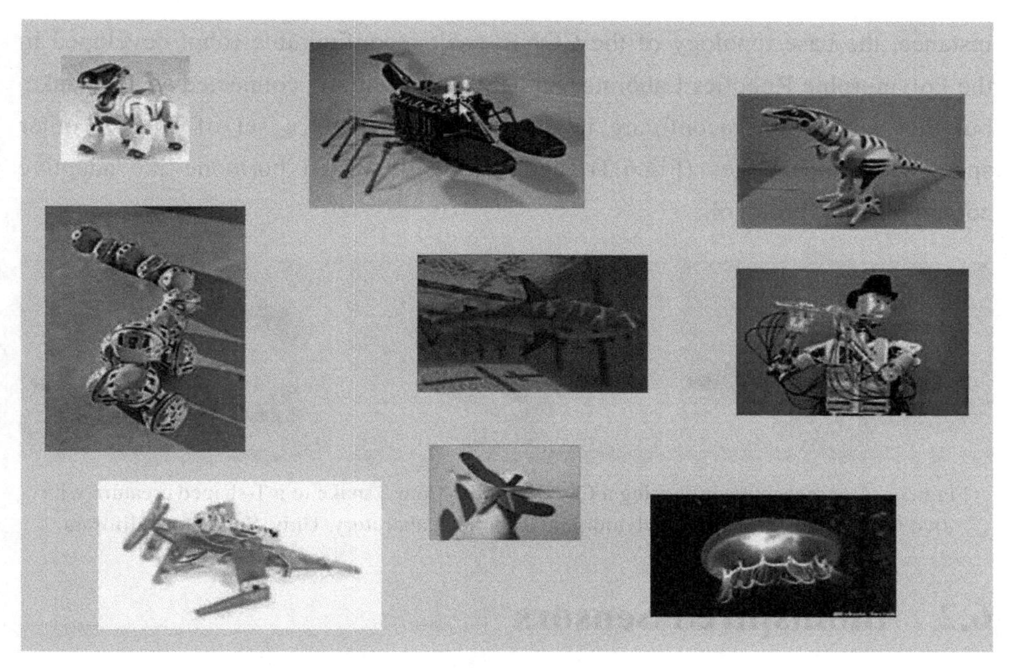

Fig.6.1 A collection of zoomorphic robots

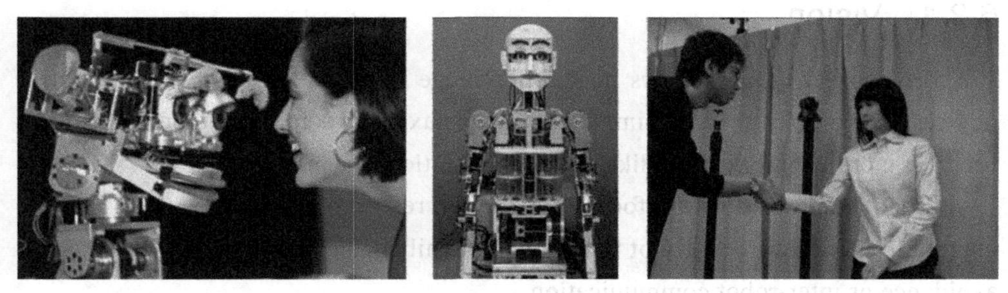

Fig.6.2 Kismet (Left), WE-4RII (Middle) and Uando (Right) humanoid robots

Likewise, the Uando robot of Osaka University is controlled by air actuators providing 43 degrees of freedom. The android can make facial expressions, eye, head, and body movements, and gestures with its arms and hands. Touch sensors with sensitivity to variable pressures are mounted under its clothing and silicone skin, while floor sensors and omnidirectional vision sensors serve to recognize where people are in order to make eye contact while addressing them during conversation. Moreover, it can respond to the content and prosody of a human partner by varying what it says and the pitch of its voice.

Another active research area in which functional considerations play a major role is that of shape-shifting robots that can dynamically reconfigure their morphology according to internal or external circumstances. Biological inspiration stems from organisms that can regrow lost appendages, like the tail in lizards, or from transitions in developmental stages, like morphogenetic changes in batrachians. For instance, the base topology of the CONRO self-reconfigurable robot developed in the Polymorphic Robotics Laboratory at USC-ISI is simply connected as in a snake, but the system can reconfigure itself in order to grow a set of legs or other specialized appendages (Fig.6.3) thanks to a dedicated hormone-like adaptive communication protocol.

 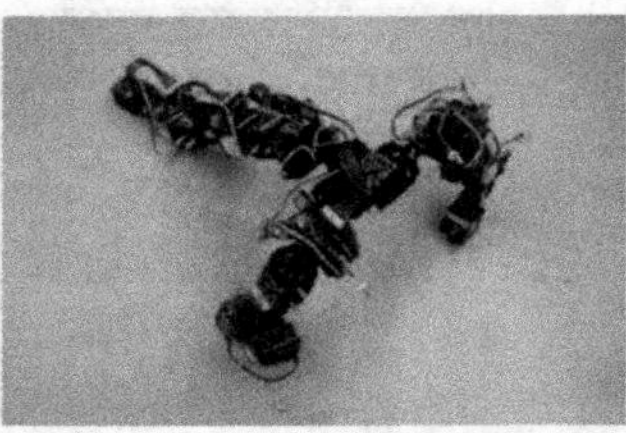 

Fig.6.3 A sequence reconfiguring a CONRO robot from a snake to a T-shaped creature with two legs. © Wei-Min Shen, Polymorphic Robotics Laboratory, Univ. Southern California

6.2 Bioinspired Sensors

6.2.1 Vision

Bioinspired visual sensors in robotics range from very simple photo sensitive devices that mostly serve to implement phototaxis, to complex binocular devices used for more cognitive tasks like object recognition, for instance.

Phototaxis is seldom the focus of dedicated research. It is rather usually implemented to merely force a robot to move and exhibit other capacities, like obstacle-avoidance or inter-robot communication.

Several visual systems calling upon optic-flow monitoring are particularly useful in the context of navigation tasks and are implemented on a variety of robots. This is the case with the work done in Marseilles' Biorobotics Laboratory that serves to understand how the organization of the compound eye of the housefly, and how the neural processing of visual information obtained during the flight, endow this insect with various reflexes mandatory for its survival. The biological knowledge thus acquired was exploited to implement opto-electronic devices allowing a terrestrial robot to wander in its environments while avoiding obstacles, or tethered aerial robots to track a contrasting target or to automatically perform terrain-following, take-off or landing (Fig.6.4).

The desert ant *Cataglyphis,* while probably merging optic-flow and odometry monitoring to evaluate its travel distances, is able to use its compound eyes to perceive the polarization pattern of the sky and infer its orientation. This affords it accurate navigation capacities that make it possible to explore its desert habitat for hundreds of meters while foraging, and return back to its nest on an almost straight line, despite the absence of conspicuous landmarks and despite the impossibility of laying pheromones on the ground that would not almost immediately evaporate. Inspired by the insect's navigation system, mechanisms for path integration and visual piloting have been successfully employed on mobile robot navigation in the Sahara desert.

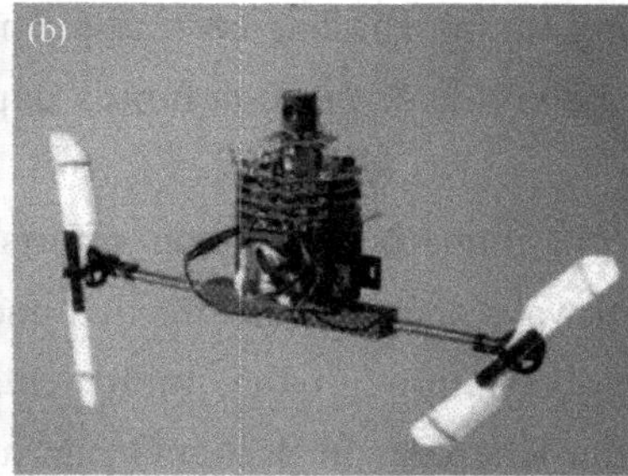

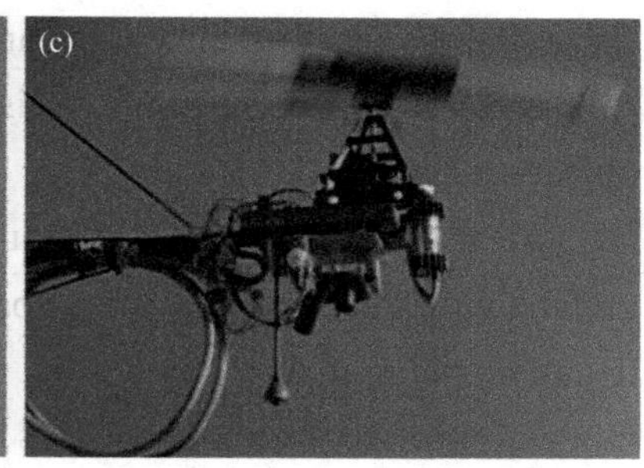

Fig.6.4 Opto-electronic devices inspired by the housefly's compound eye. (a) Device for obstacle avoidance. (b) Device for target tracking. (c) Device for terrain following, take-off, and landing. © CNRS Photothèque, Nicolas Franceschini, UMR6152 - Mouvement Perception - Marseille

Among the robotic realizations that are targeted at the humanoid vision, some aim at integrating the information provided by foveal and peripheral cameras. Some researchers, in particular, describe a system that uses shape and color to detect and pursue objects through peripheral vision and, then, recognizes the object through a more detailed analysis of higher resolution foveal images. The classification is inferred from a video stream rather than from a single image and, when the desired object is recognized, the robot reaches for it and ignores other objects. Common alternatives to the use of two cameras per eye consist in using space-variant vision and, in particular, log-polar images. As an example, an attentional system was described, which should be extended with modules for object recognition, trajectory tracking, and naive physics understanding during the natural interaction of the robot with the environment.

Vision-based SLAM (Simultaneous Localization and Mapping) systems have also been implemented on humanoid robots, with the aim of increasing the autonomy of these machines. In particular, the HRP2 robot was used to demonstrate real-time SLAM capacities during agile combinations of walking and turning motions, using the robot's internal inertial sensors to monitor a type of 3D odometry that reduced the local rate of increase in uncertainty within the SLAM map. The authors speculate that the availability of traditional odometry on all of the robot's degrees of freedom will allow more long-term motion constraints to be imposed and exploited by the SLAM algorithm, based on knowledge of possible robot configurations. As another step towards autonomy in humanoid robots, mapping and planning capacities may be combined. For instance, a real-time vision-based sensing system was demonstrated and an adaptive footstep planner allows a Honda ASIMO robot to autonomously traverse dynamic environments containing unpredictably moving obstacles.

6.2.2 Audition

Like vision, the sense of hearing in animals has been implemented on several robots to exhibit mere phonotaxis behavior or more complex capacities, such as object-recognition.

At the University of Edinburgh, numerous research efforts are devoted to understanding the sensory-motor pathways and mechanisms that underlie positive or negative phonotaxis behavior in crickets through the implementation of various models on diverse robots such as the Khepera. In particular, an analog very large scale integrated (aVLSI) circuit modeling the auditory mechanism that serves a female cricket to meet a conspecific male or to evade a bat—by the calling song or the echolocation calls they respectively produce—has been built. The corresponding results suggest that the mechanism outputs a directional signal to sounds ahead at calling song frequency and to sounds behind at echolocation frequencies, and that this combination of responses simplifies later neural processing in the cricket. This processing is the subject of complementary modeling efforts in which spiking neuron controllers are also tested on robots, thus allowing exploring the functionality of identified neurons in the insect, including the possible roles of multiple sensory fibers, mutually inhibitory connections, and brain neurons with pattern-filtering properties. Such robotic implementations also make the investigation of multimodal influences on the behavior possible, via the inclusion of an optomotor stabilization response and the demonstration that may improve auditory tracking, particularly under conditions of random disturbance.

Concerning more cognitive capacities, within the framework of the EC (European Community) project CIRCE (Chiroptera Inspired Robotic CEphaloid), a bat head is used to investigate how the world is not just perceived, but actively explored, by bats. In particular, the work aims at identifying how the various shapes, sizes, and movements influence the signals the animal receives from its environment. It is hoped that the principles gleaned from such work will prove useful in developing better antennas, particularly for wireless devices that are in motion and need to pick up complex signals from different directions.

Likewise, the Yale Sonar Robot that is modeled after bat and dolphin echolocation behavior is said to be so sensitive that it can tell whether a tossed coin has come up heads or tails. Called Rodolph—short for Robotic dolphin, the robot is equipped with electrostatic transducers that can act either as transmitters or receivers to serve as the robot's "mouth" and "ears". The design is inspired by bats, whose ears react by rotating in the direction of an echo source, and by dolphins, which appear to move around in order to place an object at a standard distance, thus reducing the complexity of object recognition.

Some researchers describe a system that allows a humanoid robot to listen to a specific sound source under noisy environments—a human capability that is known as "cocktail party effect"—and to listen to several speeches simultaneously, thus allowing to cope with situations where someone or something playing sounds interrupts conversation—a capacity known as "barge-in" in spoken dialog systems. This system calls upon active motions directed at the sound source to improve localization by exploiting an "auditory fovea". It also capitalizes on audio-visual integration, thus making localization, separation and recognition of three simultaneous speeches possible.

6.2.3 Touch

It is often asserted that, of all the five senses, touch is the most difficult to replicate in mechanical form. Be that as it may, a passive, highly compliant tactile sensor has been designed for the hexa-pedal running robot Sprawlette at Stanford, drawing inspiration by how the cockroach *Periplaneta americana* uses antenna feedback to control its orientation during a rapid wall-following behavior. Results on the stabilization of the robot suggest that the cockroach uses, at least in part, the rate of convergence to the wall—or "tactile flow"—to control its body orientation. To make it possible to detect the point of greatest strain, or to differentiate between different shapes the sensor is bent into, more advanced versions of the antenna are currently under development.

While a cockroach's antenna consists of multiple rigid segments and is covered all along its length with sensory receptors, a rat's whisker consists of a single, flexible, tapered hair and has tactile sensors located only at its base. The way two arrays of such sensors afford capacities of obstacle-avoidance, texture discrimination and object recognition, has inspired several robotic realizations notably that one in which the whiskers are passive and rely upon the motion of the robot in order to scan the surface profile of touched objects. The robot is able to recognize a few objects formed from plane, cylindrical and spherical surfaces. By using its simple manipulator, it can pick up and retrieve small objects.

Conversely, a touch system was described based on computational models of whisker-related neural circuitry in the rat brain, in which the whiskers will be actively scanning the surroundings. This work will contribute to the EC project ICEA, whose primary aim is to develop a cognitive systems architecture integrating cognitive, emotional and bioregulatory (self-maintenance) processes, based on the architecture and physiology of the mammalian brain.

In the field of humanoid robotics, investigations on touch sensors are conducted at the University of Tokyo, where a robotic hand calling upon organic transistors as pressure sensors [Fig.6.5(a)] has been produced. The same technology served to make a flexible artificial skin that can sense both pressure and temperature [Fig.6.5(b)], thus more closely imitating the human sense of touch.

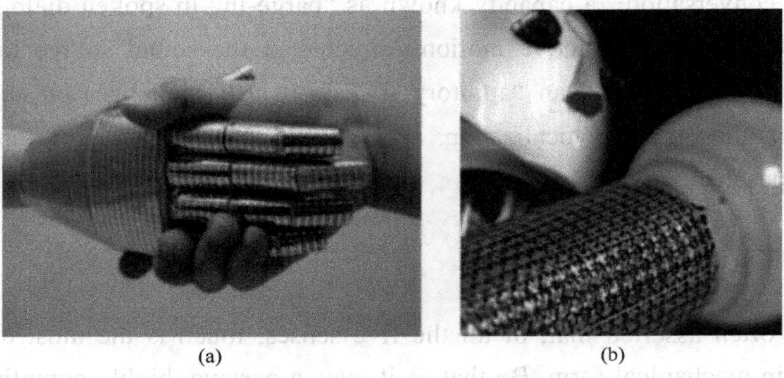

(a) (b)

Fig.6.5 Artificial skin devices at Tokyo University. (a) Pressure-detection. (b) Pressure and temperature detection

Another step in this direction has been made at the University of Nebraska where a thin-film tactile sensor, which is as sensitive as the human finger in some ways, has been designed. When pressed against a textured object, the film creates a topographical map of the surface, by sending out both an electrical signal and a visual signal that can be read with a small camera. The spatial resolution of these "maps" is as good as that achieved by human touch, as demonstrated by the image obtained when putting a penny on this mechanical fingertip. Although such sensor deals with texture in a way that is not at all like a fingertip, it has a high enough resolution to "feel" single cells, and therefore could help surgeons find the perimeter of a tumor during surgical procedures. Cancer cells—in particular, breast cancer cells—have levels of pressure that are different from normal cells, and should feel "harder" to the sensor.

6.2.4 Smell

The way the nematod *Caenorhabditis elegans* uses chemotaxis—probably the most widespread form of goal-seeking behavior—to find bacterial food sources by following their odors has been investigated at the University of Oregon. The worm has a small nervous system (302 neurons), whose neurons and connectivity pattern have been completely characterized, the neural circuit controlling chemotaxis is well known and, when implemented on a robot, it proves to be able to cope with environmental variability and noise in sensory inputs. The long-term objective of such work is to design a cheap, artificial eel that could locate explosive mines at sea. Among the research efforts that tackle the related and highly challenging issue of reproducing the odor plume-tracking behavior in marine animals, recent results obtained on the RoboLobster are put in perspective.

Other bioinspired systems for odor recognition are under development in several places. For instance, the chest of the humanoid WE-4RII robot of Waseda University is equipped with two mechanical lungs each consisting of a cylinder and a piston, thanks to which the robot breathes air. Being also equipped with four semiconductor gas sensors, it recognizes the smells of alcohol, ammonia and cigarette smoke.

6.2.5 Taste

A first robot with a sense of taste has been recently developed by NEC System Technologies, Ltd. Using infrared spectroscopic technology, this robot is capable of examining the taste of food and giving its name as well as its ingredients. Furthermore, it can give advice on the food and on health issues based on the information gathered. The latest developments afford the robot with capacities of distinguishing good wine from bad wine, and Camembert from Gouda.

Whereas the preceding sensors were all providing information about an animal's or a robot's external world—they are called allothetic sensors by biologists, other sensors may provide information about a creature's internal state. Although such so-called idiothetic sensors are widespread in robotic applications—measuring variables like temperature, pressure, voltage, accelerations, etc., they are seldom biologically inspired, but in the implementation of a variety of visual-motor routines (smooth-pursuit tracking, saccades, binocular vergence, vestibular-ocular and opto-kinetic reflexes), like those that are at work in the humanoid Cog robot mentioned later.

6.3 Bioinspired Actuators

(1) Crawling

Because they are able to move in environments inaccessible to humans, such as pipes or collapsed buildings, numerous snake-like robots have been developed for exploration and inspection tasks, as well as for participation to search and rescue missions. The AmphiBot of EPFL extends the capacities of these robots because it is amphibious and capable of both swimming and lateral undulatory locomotion. Being inspired by central pattern generators (CPG) found in vertebrate spinal cords, it also contributes to better understand how their central nervous system controls movement in animals like snakes and elongate fishes such as lampreys.

Other applications are sought within the framework of the EC project BIOLOCH (BIO-mimetic structures for LOComotion in the Human body). In the perspective of helping doctors diagnose disease by carrying tiny cameras through patients' bodies, a robot designed to crawl through the human gut by mimicking the wriggling motion of an undersea worm has been developed by the project partners. Drawing inspiration from the way polychaetes, or "paddle worms", use tiny paddles on their body segments to push through sand, mud or water, they tackled the issue of supplying traditional forms of robotic locomotion that would not work in the peculiar environment of the gut. The device is expected to lessen the chance of damaging a patient's internal organs with a colonic endoscope, and to enhance the exploration capacities afforded by "camera pills".

(2) Walking

In the PolyPEDAL (Performance Energetics and Dynamics of Animal Locomotion) Laboratory at Berkeley, general principles about legged locomotion are sought, through the comparison of the sensory-motor equipment and the behavior of a variety of animals. In particular, it has been discovered that many animals self-stabilize to perturbations without a brain or its equivalent because control algorithms are embedded in their physical structure. Shape deposition manufacturing has allowed engineers to tune legs of the SPRAWL family of hand-sized hexa-pedal robots inspired by the cockroach that are very fast (up to 5 body-lengths per second), robust (hip-height obstacles), and that self-stabilize to perturbations without any active sensing. A cricket-inspired robot, approximately 8 cm long, designed for both walking and jumping is under development at Case Western Reserve University. McKibben artificial muscles will actuate the legs, compressed air will be generated by an onboard power plant, and a continuous-time recurrent neural network will be used for control. Additionally, front legs will enable climbing over larger obstacles and will also be used to control the pitch of the body before a jump and, therefore, aim at the jump for distance or height.

Engineers from Boston Dynamics claim they have developed "the most advanced quadruped robot on Earth" for the US Army. Called BigDog, it walks, runs, climbs on rough terrain, and carries heavy loads. Being the size of a large dog or a small mule, measuring 1 m long, 0.7 m tall and 75 kg weight, BigDog has trotted at 5 km/h, climbed a 35° slope, and carried a 50 kg load so far. BigDog is powered by a gasoline engine that drives a hydraulic actuation system. Its legs are articulated like an animal's, and have compliant elements that absorb shock and recycle energy from one step to the next. Another quadruped with amazing locomotion capabilities is Scout Ⅱ, presumably the world's first galloping robot, developed at McGill University. Using a single actuator per leg—the hip joint providing leg rotation in the sagittal plane, and each leg having two degrees of freedom (DOF)—the actuated revolute hip DOF, and the passive linear compliant leg DOF, the system exhibits passively generated bounding cycles and can stabilize itself without the need of any control action. This feature makes simple open-loop control of complex running behaviors such as bounding and galloping possible.

Developed at Stirling University, RunBot is probably the world's fastest biped robot for its size. Being 30 cm high, it can walk at a speed of 3.5 leg-lengths per second, which is comparable to the fastest relative speed of human walking. This robot has some special mechanical features, e.g., small curved feet allowing rolling action and a properly positioned center of mass, that facilitate fast walking through exploitation of its natural dynamics. It also calls upon a sensor-driven controller that is built with biologically inspired sensor and motor-neuron models, the parameters of which being possibly tuned by a policy gradient reinforcement learning algorithm in real-time during walking. The robot does not employ any kind of position or trajectory-tracking control algorithm. Instead, it exploits its own natural dynamics during critical stages of its walking gait cycle.

(3) Wall-climbing

In the Biomimetic Dextrous Manipulation Laboratory at Stanford University, researchers are working on a gecko-like robot, called Stickybot, designed to climb smooth surfaces like glass without using suction or adhesives (Fig.6.6). Geckos can climb up walls and across ceilings thanks to roughly half a million of tiny hairs, or setae, on the surface of each of their feet and to the hundreds to thousands of tiny pads, or spatulae, at the tip of each hair. Each of these pads is attracted to the wall by intermolecular van-der-Waals forces, and this allows the gecko's feet to adhere. Conversely, if the hair is levered upward at a 30 degree angle, the spatulae at the end of the hair easily detach. The gecko does this simply by peeling its toes off the surface. Inspired by such structures and mechanisms, the Stickybot's feet are covered with thousands of synthetic setae made of an elastomer. These tiny polymer pads ensure a large area of contact between the feet and the wall, thus maximizing the expression of intermolecular forces. In the same laboratory, a six-legged robot called Spinybot climbs vertical surfaces according to similar principles. Spinybot's feet and toes are made from several different polymers, which range from flexible to rigid, thus enabling the robot to absorb jolts and bumps, much as animals' feet do.

The project RiSE (Robots in Scansorial Environments) funded by the DARPA Biodynotics Program constitutes an extension of these research efforts that aims at building a bioinspired climbing robot with the unique ability to walk on land and climb on trees, fences, walls, as well as other vertical surfaces. It calls upon novel robot kinematics, precision-manufactured compliant feet and appendages, and advanced robot behaviors (Fig.6.6).

Fig.6.6 Wall climbing robots at Stanford. Left : Stickybot. Right : RiSE robot

(4) Jumping

In the perspective of environment exploration and monitoring, some researchers described a lightweight micro-robot that demonstrates that jumping can be more energetically efficient than just walking or climbing, and which can be used to overcome obstacles and uneven terrains. During the flight phase, energy from an electric micro-motor is collected in the robot's springs, while it is released by a click mechanism during take-off. In this way, instant power delivered by rear legs is much higher than the one provided by the motor.

(5) Swimming

Several biomimetic robots are being produced that emulate the propulsive systems of fish, dolphins, or seals, and exploit the complex fluid mechanics these animals use to propel themselves. A primary goal of these projects is to have machines that can maneuver by taking advantage of flows and body positions, leading to huge energy savings, and substantially increasing the length of swimming time. For instance, the group at MIT Towing Tank made two robotic fish, a "RoboTuna" and a "RoboPike", that use servo motors and spring element spines (Fig.6.7), and serve to demonstrate the advantages of flapping foil propulsion. It has thus been shown that RoboTuna can reduce its drag in excess of 70% compared to the same body towed straight and rigid. Likewise, it appears that biomimetic fish can turn at a maximum rate of 75 (°)/s, whereas conventional rigid-bodied robots and submarines turn at approximately 3-5 (°)/s.

Fig.6.7 MIT swimming robots. Left: RoboTuna. Right: RoboPike

The robot Madeleine of Vassar College imitates the design of a turtle. It has a comparable power output, and its polyurethane flippers have the same stiffness as a real turtle's, but are operated by electric motors connected to an onboard computer. Because it may swim underwater using four flippers, like many extinct animals, or with two flippers, like modern animals, this robot has been used to test theories of locomotion in existing and extinct animals. It thus appears that having four flippers does not improve the top speed—apparently because the front flippers created turbulence that interfered with the rear flippers' ability to generate forward propulsion—but does increase energy use. This may explain why natural selection favored two-flipper animals over four-flipper animals like the plesiosaurs, and why four-flipper animals such as penguins, sea turtles and seals use only two of their limbs for propulsion.

(6) Flying

Flapping wings offer several advantages over the fixed wings of today's reconnaissance drones, like flying at low speeds, hovering, making sharp turns and even flying backward. Like in animals, the vortex created beneath each wing is exploited to create the push necessary for robots to take to the sky.

The goal of the Micromechanical Flying Insect (MFI) project at Berkeley is to develop a 25 mm robot capable of sustained autonomous flight, which could be used in search, rescue, monitoring, and reconnaissance. Such tiny robot will be based on biomimetic principles that capture some of the exceptional flight performance achieved by true flies, i.e., large forces generated by non-steady state aerodynamics, a high power-to-weight ratio motor system, and a high speed control system with tightly integrated visual and inertial sensors. Design analysis suggests that piezoelectric actuators and flexible thorax structures can provide the needed power density and wing stroke, and that adequate power can be supplied by lithium batteries charged by solar cells. Likewise, mathematical models capitalizing on wing-thorax dynamics, flapping flight aerodynamics at a low Reynolds number regime, body dynamics, as well as on a biomimetic sensory system consisting of ocelli, halteres, magnetic compass, and optical flow sensors, have been used to generate realistic simulations for MFI and insect flight. In turn, such simulations served to design a flight control algorithm maintaining a stable flight in hovering mode. A first MFI platform, which flaps its two wings and is the right size, has already been produced.

The four-winged ornithopter Mentor, which is developed at the University of Toronto as part of a general research effort targeted at flapping-wing flight, is said to be the first artificial device that successfully hovered, doing so with the agility of a hummingbird. In particular, it exhibited the "clap-fling" behavior that the animal uses to draw in air by clapping its wings together, then flinging them apart at high speeds. This creates lift by hurling regions of high pressure below and behind. Likewise, the way it elegantly shifts from hovering to horizontal flight inspires current research. Mentor is about 30 cm long and weighs about 0.5 kg, but engineers hope to eventually shrink it to hummingbird size and weight. Other comparable MAV (Micro Aerial Vehicle) devices are reported.

On a much larger scale, a few manned flapping-wing robots have also been designed. In Votkinsk in the 1990's, Toporov built a biplane tow-launched ornithopter that reportedly could be made to climb and fly for 200 m as a result of the pilot's muscular effort. More recently, within the Ornithopter project of SRI International and University of Toronto, the two-winged Flapper plane has flown for 14 sec at an average speed of 88 km/h. It has a 12 m wingspan and weighs 350 kg with pilot and fuel. The wings are made of carbon fiber and Kevlar, and are moved by a gas-powered engine.

Finally, despite general skepticism, there are plans for commercial applications of flapping-wing flight, and a prototype with 16 wings and 125 seats is announced to be under development at JCR Technology Corporation.

(7) Grasping

When hunting and grabbing food, the octopus uses all the flexibility its arms are capable of. But, when feeding, the animal is able to bend its flexible arms to form "joints" like those in human arms. Inspired by such dexterous appendages found in cephalopods—particularly the arms and suckers of octopus, and the arms and tentacles of squid, researchers describe recent results in the development of a new class of soft, continuous backbone robot manipulators. Fed by fundamental research into the manipulation tactics, sensory biology, and neural control of octopuses, the work in turn leads to the development of artificial devices based on both electro-active polymers and pneumatic McKibben muscles, as well as to novel approaches to motion planning and operator interfaces for the so-called OCTARM robot. Likewise, inspired by biological trunks and tentacles, a multi-section continuum robot, Air-Octor, in which the extension of each section can be independently controlled, exhibits both bending and extension capacities, and demonstrates superior performance arising from the additional degrees of freedom than arms with comparable total degrees of freedom.

Human grasping has inspired the humanoid hand developed at Curtin University of Technology. The corresponding system presents 10 individually controllable degrees of freedom ranging from the elbow to the fingers, and is actuated through 20 McKibben air muscles each supplied by a pneumatic pressure-balancing valve that allows for proportional control to be achieved with simple and inexpensive components. The hand is able to perform a number of human-equivalent tasks, such as grasping and relocating objects (Fig.6.8). A similar research is funded by the EC CYBERHAND project that aims at developing a cybernetic prosthetic hand. It is hoped that the device will re-create the "life-like" perception of the natural hand, and thus increase its acceptability. To this end, biomimetic sensors replicating the natural sensors are to be developed, and dedicated electrodes—capable of delivering sensory feedback to the amputee's central nervous system and extracting his intentions—are to be designed (Fig.6.8).

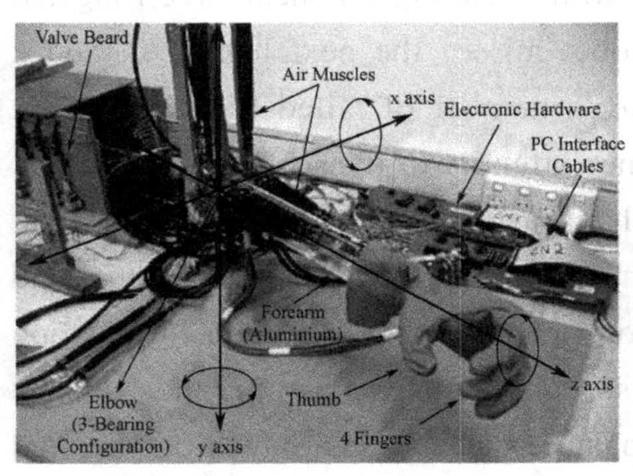

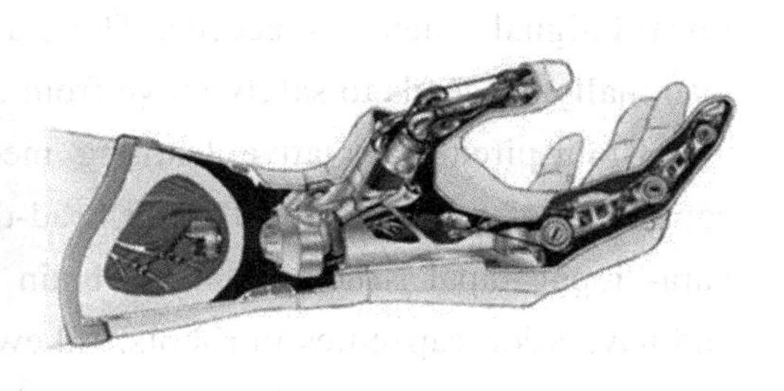

Fig.6.8 Left: The humanoid hand of Curtin University of Technology.
Right: The CYBERHAND project

6.4 Bioinspired Control Architectures

In this section, more cognitive architectures, able to deal with past and future events as well, and in which adaptive mechanisms like learning, evolution and development may be incorporated, will be mentioned.

6.4.1 Behavior-based Robotics

Under the aegis of so-called behavior-based robotics, many systems with minimally-cognitive architectures have been developed. For instance, the series of robots designed by Brooks and his students at MIT demonstrate that the "subsumption architecture" may endow artificial animals with adaptive capacities that do not necessitate high-level reasoning. Moreover, there are some indications that such control architecture may be at work in real animals, like the coastal snail Littorina for example. Likewise, the "schemas" that are used by Arkin and his students at the Georgia Institute of Technology to control numerous other robots have roots in psychology and neuroscience.

6.4.2 Learning Robotics

Different bioinspired learning mechanisms—like those implementing associative, reinforcement or imitation learning schemes—are currently at work in robotic applications.

For instance, in the robotics laboratory at Nagoya University, the robot Brachiator is able to swing from handhold to handhold like a gibbon. The robot is equipped with legs that generate initial momentum, and with a computer vision system to figure out where to place its hand-like grippers. A standard reinforcement learning algorithm is used to learn the right sensory-motor coordination required to move along a horizontal scale while hanging on successive rungs: it provides a punishment signal when the robot misses the next handhole, and a reward signal when it succeeds. Thus, after a number of failed trials, the robot eventually succeeds to safely move from one extremity of the scale to the other.

Bioinspired associative learning mechanisms are used in applications that capitalize upon the place cells and head-direction cells found in hippocampal and para-hippocampal structures in the brain to implement map-building, localization and navigation capacities in robots. Likewise, reinforcement learning mechanisms inspired by the presumed function of dopaminergic neurons may be associated with models based on the anatomy and physiology of basal ganglia and related structures, which endow a robot with a motivational system and action-selection capacities—i.e., those of deciding when to shift from one activity to another, according to the various sub-goals the surprises encountered during the fulfillment of a given mission generate. Such controllers and capacities are currently combined in the Psikharpax artificial rat—that will be able to explore an unknown environment, to build a topological map of it, and to plan trajectories to places where it will fulfill various internal needs, like “eating”, “resting”, “exploring” or “avoiding danger”—as a contribution to the EC project ICEA mentioned before.

MirrorBot, another EC project, capitalizes on the discovery of mirror neurons in the frontal lobes of monkeys, and on their potential relevance to human brain evolution. Indeed, mirror neuron areas correspond to cortical areas which are related to human language centers, and it seems that these neurons have a critical role in cortical networks establishing links between perception, action and language. The project has developed an approach of biomimetic multimodal learning, including imitation learning, using a mirror-neuron-based robot, and has investigated the task of foraging for objects that are designed by their names.

At the Neuroscience Institute in San Diego, a series of brain-based devices (BBDs) —i.e. physical devices with simulated nervous systems that guide behavior, to serve as a heuristic for understanding brain function—have been constructed. These BBDs are based on biological principles and alter their behavior to the environment through self-learning. The resulting systems autonomously generalize signals from the environment into perceptual categories and through adaptive behavior, become increasingly successful in coping with the environment. Among these devices, the robot Darwin Ⅶ is equipped with a CCD camera for vision, microphones for hearing, conductivity sensors for taste, and effectors to move its base and its head, and with a gripping manipulator having one degree-of-freedom. Its control architecture is made of 20,000 brain cells, and it is endowed with a few instincts, like an interest in bright objects, a predilection for tasting things, and an innate notion of what tastes good. Thus, the robot explores its environment and quickly learns that striped blocks are yummy and that spotted ones taste bad. Based on the same robotic platform, Darwin Ⅷ is equipped with a simulated nervous system containing 28 neural areas, 53,450 neuronal units, and approximately 1.7 million synaptic connections. It demonstrates that different brain areas and modalities can yield a coherent perceptual response in the absence of any superordinate control, thus solving the so-called binding problem. In particular, the robot binds features such as colors and line segments into objects and discriminates between these objects in a visual scene. Darwin Ⅸ is a mobile physical device equipped with artificial whiskers and a neural simulation based on the rat somatosensory system. Neuronal units with time-lagged response properties, together with the selective modulation of neural connection strengths, provide a plausible neural mechanism for the spatiotemporal transformations of sensory input necessary for both texture discrimination and selective conditioning to textures. Having an innate tendency to avoid "foot-shock" pads made of reflective construction paper deposited on the ground of its experimental arena, the robot may be conditioned to avoid specific textures encountered near these aversive stimuli. Darwin Ⅹ incorporates a large-scale simulation of the hippocampus and surrounding areas, thus making it possible to solve a dry version of the Morris "water-maze" task, in which the robot must find a hidden platform in its environment using only visual landmarks and self-movement cues to navigate to the platform from any starting position. Besides its ability to learn to run mazes like rats, Darwin Ⅹ has been thrown in a soccer match, and turned out to be victorious in the 2005 RoboCup U.S. Open. Finally, Darwin Ⅺ combines the main characteristics of several previous versions, including a whisker system, and serves to demonstrate the robot's capacity to learn the reward structure of the environment, as well as the reversal of behavior when this structure changes.

In the perspective of exploring the role that chaotic dynamics may play in self-organizing behavior, researchers involved in the CICT-funded SODAS (Self-Organizing Dynamically Adaptable Systems) project are using a nonlinear dynamics approach to model how the brain—which is usually in a high-dimensional, disorderly "basal" state—instantly shifts from a chaotic state to an attractor four or five times a second in order to recognize something familiar, or to make a decision. Such phase transitions and attractors in one area of the brain affect attractors in other areas, and are considered to produce intentional behavior. Focused on the way the brain orients the body in space and uses positive and negative reinforcement from the environment to autonomously navigate to a destination, the goal of the SODAS project is to enable robots to do the same on future NASA missions. In particular, it has produced the KIV architecture that models the brain's limbic system, the simplest neurological structure capable of acting intentionally in an inherently self-consistent manner. Some researchers describe how, in a 2D computer simulation of a Martian landscape, KIV uses positive and negative reinforcement to learn the most effective path to a goal, and uses habituation to reduce the distraction of ambient noise and other irrelevant sensory inputs.

Key vocabularies

actuator [ˈæktʃʊeɪtə] *n.* 激励者；促动器，执行机构驱动器，执行器

aegis [ˈiːdʒɪs] *n.* 保护；支持；由……主办；在……的支持下

agile [ˈædʒaɪl] ***adj.*** 灵巧的；轻快的；灵活的；机敏的
allegedly [əˈledʒɪdlɪ] ***adv.*** 据说，据称
allothetic ***adj.*** 异体的
amphibious [æmˈfɪbiəs] ***adj.*** 两栖的；水陆两用的
amputee [ˌæmpjuˈtiː] ***n.*** 被截肢者
animatronics [ˌænɪməˈtrɒnɪks] ***n.*** 电子动画学，电动木偶技术
antenna [ænˈtenə] ***n.*** 触须，触角；天线；感觉，直觉
anthropopathic [ˈænθrəpəpəθɪk] ***adj.*** 拟人的；认为神具有人性的说法的
auditory [ˈɔːdətri] ***adj.*** 听觉的，听力的
autonomy [ɔːˈtɒnəmi] ***n.*** 自治，自治权；独立自主，自主权
basal ganglia ***n.*** 基底神经节；基底核
batrachians [bəˈtreɪkɪən] ***n.*** 无尾两栖类动物，蛙或蟾蜍. ***adj.*** 无尾两栖类动物的
binocular [bɪˈnɒkjələ(r)] ***adj.*** 双眼的；双目并用的
biped [ˈbaɪped] ***adj.*** 有两足的. ***n.*** 两足动物
biplane [ˈbaɪpleɪn] ***n.*** 双翼飞机，复翼飞机
calling song ***n.*** 呼唤性鸣声
capitalize [ˈkæpɪtəlaɪz] ***v.*** 利用
cephalopod [ˈsefələpɒd] ***n.*** 头足纲动物（如章鱼）
chemotaxis [keməˈtæksɪs] ***n.*** 趋化作用；趋化现象；趋药性；趋化性
circa [ˈsɜːkə] ***prep.*** 大约，左右（主要用于日期前）. ***adv.*** 大约
coherent [kəʊˈhɪərənt] ***adj.*** 有条理的，连贯的；说话条理清晰的，易于理解的
colonic [kəˈlɒnɪk] ***adj.*** 结肠的；灌洗结肠的
conspecific [ˌkɒnspɪˈsɪfɪk] ***adj.*** （植物、动物）同种的
cricket [ˈkrɪkɪt] ***n.*** 蟋蟀；板球（运动）
cybernetic [ˌsaɪbəˈnetɪk] ***adj.*** 控制论的
dexterous [ˈdekstrəs] ***adj.*** （身手）灵巧的，敏捷的
dopaminergic [dəʊpəmɪˈnɜːdʒɪk] ***adj.*** 多巴胺能的
drone [drəʊn] ***n.*** 无人驾驶飞机（或导弹）
echolocation [ˌekəʊləʊˈkeɪʃn] ***n.*** 回波定位；回声测距；回声定位能力
eel [iːl] ***n.*** 鳝鱼；鳗鱼；小线虫；精明油滑的人
elastomer [ɪˈlæstəmə(r)] ***n.*** 弹性体，高弹体，人造橡胶
endoscope [ˈendəskəʊp] ***n.*** 内窥镜；内诊镜
flipper [ˈflɪpə(r)] ***n.*** 鳍状肢；鳍
foil [fɔɪl] ***n.*** 箔；箔纸
foveal ***adj.*** 中心凹的
gibbon [ˈɡɪbən] ***n.*** 长臂猿
glean [ɡliːn] ***v.*** 收集（资料）；拾（落穗）
gripper [ˈɡrɪpə] ***n.*** 夹子，钳子；抓器，抓爪

gut [gʌt] ***n.*** 肠，肠道；内脏
halter [ˈhɔːltə(r)] ***n.*** 缰绳；绞索．***vt.*** 给……套上缰绳；束缚
handhole [ˈhændhəʊl] ***n.*** （筛眼）手孔，探孔
heuristic [hjuˈrɪstɪk] ***n.*** 启发式步骤（或方法）
hippocampal [ˌhɪpəˈkæmpəl] ***adj.*** 海马的；海马趾的
humanoid [ˈhjuːmənɔɪd] ***n.*** 仿真机器人；类人动物．***adj.*** 像人的
hummingbird [ˈhʌmɪŋbɜːd] ***n.*** 蜂鸟
hurl [hɜːl] ***v.*** 猛投，用力投掷；大声说出
iguana [ɪˈgwɑːnə] ***n.*** 鬣蜥蜴
jolt [dʒəʊlt] ***n.*** 颠簸，摇晃
lamprey [ˈlæmpri] ***n.*** 七鳃鳗；八目鳗
limbic [ˈlɪmbɪk] ***adj.*** 边的，缘的
lobe [ləʊb] ***n.*** 耳垂；肺叶；脑叶；裂片
mammalian [mæˈmeɪliən] ***n.*** 哺乳动物．***adj.*** 哺乳动物的
maze [meɪz] ***n.*** 迷宫；错综复杂的事物
modality [məʊˈdæləti] ***n.*** 形式，形态；程序；物理疗法；主要的感觉
morphogenetic [mɔːfəʊdʒɪˈnetɪk] ***adj.*** 有关形态发生的，有关器官形成的；地貌成因的
mule [mjuːl] ***n.*** 骡，骡子；杂交动物，杂交植物
necessitate [nəˈsesɪteɪt] ***vt.*** 使……成为必要，需要；强迫，迫使
nematod ***n.*** 线虫纲
ocelli [əʊˈselaɪ] ***n.*** 单眼，眼点（ocellus 的复数）
octopus [ˈɒktəpəs] ***n.*** 章鱼；章鱼肉
ocular [ˈɒkjələ(r)] ***adj.*** 眼睛的；适于眼睛的；用眼的；视觉的
odometry [ˈɒdɒmɪtrɪ] ***n.*** 量距；测程法
omnidirectional [ˌɒmnɪdəˈrekʃənl] ***adj.*** 全方向的；全指向式的（麦克风）
onboard power plant ***n.*** 机载发电设备，机载电站
opto-kinetic ***adj.*** 光动力的
optomotor [ɒptəˈməʊtə] ***adj.*** 视动的
ornithopter [ˌɔːnɪˈθɒptə] ***n.*** 扑翼飞机
para-hippocampal ***adj.*** 副海马区的，海马旁的
perceptual [pəˈseptʃuəl] ***adj.*** 知觉的，有知觉的；感性的
peripheral [pəˈrɪfərəl] ***adj.*** 次要的，附带的；外围的，周边的；周围神经系统的
per se [ˌpɜːˈseɪ] ***adv.*** 本身，本质上
perturbation [ˌpɜːtəˈbeɪʃn] ***n.*** 忧虑；不安；烦恼；摄动
pheromone [ˈferəməʊn] ***n.*** 信息素（用于生化领域）；外激素
phototaxis [ˌfəʊtəˈtæksɪs] ***n.*** 趋光性；慕光性
physiology [ˌfɪziˈɒlədʒi] ***n.*** 生理学；生理机能

piezoelectric [paɪˌiːzəʊɪˈlektrɪk] ***adj.*** 压电的
plesiosaur [ˈpliːsɪəsɔː] ***n.*** 蛇颈龙
polychaete [ˈpɒlɪkiːt] ***n.*** 多毛类环虫；多毛动物. ***adj.*** 多毛纲的
polyurethane [ˌpɒliˈjʊərəθeɪn] ***n.*** 聚氨基甲酸酯；聚氨酯
predilection [ˌpriːdɪˈlekʃn] ***n.*** 偏爱，偏好，嗜好
prosody [ˈprɒsədi] ***n.*** 韵文学；诗体学；（某语言的）韵律（学）
prosthetic [prɒsˈθetɪk] ***adj.*** 义肢的，假体的；非朊基的
quadruped [ˈkwɒdruped] ***n.*** 四足动物. ***adj.*** 有四足的
reconnaissance [rɪˈkɒnɪsns] ***n.*** 侦察或观测，勘查，勘探；选点
recurrent [rɪˈkʌrənt] ***adj.*** 复发的，复现的；周期性的，经常发生的；回归的；循环的
resemblance [rɪˈzembləns] ***n.*** 相似，形似；外表，外观；相似物，相似点；肖像
retrieve [rɪˈtriːv] ***v.*** 找回，收回；检索（储存于计算机的信息）
revolute [revəˈluːt] ***adj.*** [植]外卷的； [植]后旋的.
robust [rəʊˈbʌst] ***adj.*** 强健的，强壮的；（系统或组织）稳固的，健全的
saccade [sæˈkɑːd] ***n.*** 眼睛飞快扫视；急速勒马
sagittal [ˈsædʒətəl] ***adj.*** 矢状的，前后向的；（位于）矢形面的；箭样的
somatosensory [ˌsəʊmətəʊˈsensərɪ] ***adj.*** （耳、目、口等以外的）体觉的
spinal cord [ˈspaɪnəl kɔːd] ***n.*** 脊髓
squid [skwɪd] ***n.*** 枪乌贼，鱿鱼
subsumption [səbˈsʌmpʃən] ***n.*** 包容；包含；类别；小前提
sucker [ˈsʌkə(r)] ***n.*** （动物）吸盘；（橡胶）吸盘
synaptic [sɪˈnæptɪk] ***adj.*** 突触的；与突触有关的；（染色体）接合的
synergy [ˈsɪnədʒi] ***n.*** 协同增效作用；协同，配合
tactic [ˈtæktɪk] ***n.*** 策略，手法. ***adj.*** 按顺序的，依次排列的
tactile [ˈtæktaɪl] ***adj.*** 触觉的；触觉感知的；能触知的；有形的
tentacle [ˈtentəkl] ***n.*** 触手；触角；触须
tether [ˈteðə(r)] ***n.*** 系绳；限度，极限. ***vt.*** 用绳拴住；拘束，束缚
thorax [ˈθɔːræks] ***n.*** 胸；胸部；胸腔
topology [təˈpɒlədʒɪ] ***n.*** 地质学；拓扑结构；局部解剖学
tow-launched ***adj.*** 牵引式启动的，牵引式发射的
transducer [trænzˈdjuːsə(r)] ***n.*** 传感器，变频器，变换器
trot [trɒt] ***n.*** （马等动物的）小跑，慢跑. ***v.*** 快步，疾走；步行
undulatory [ˈʌndjʊlətərɪ] ***adj.*** 波动的，起伏的
vergence [ˈvɜːdʒəns] ***n.*** 趋异；聚散度；朝向
vestibular [vesˈtɪbjʊlə] ***adj.*** 门厅的，门口走廊的，前庭的
wasp [wɒsp] ***n.*** 黄蜂；胡蜂
whisker [ˈwɪskə(r)] ***n.*** 细须；连鬓胡子；须晶

Exercises

Ⅰ. Translate the following English into Chinese.

1. As such qualities are required in a continually growing application field—from planetary exploration to domestic uses, a spectacular reversal of interest towards living creatures can be noticed in current-day robotics, up to the point that it has been said that natural inspiration is the "new wave" of robotics.

2. Touch sensors with sensitivity to variable pressures are mounted under its clothing and silicone skin, while floor sensors and omnidirectional vision sensors serve to recognize where people are in order to make eye contact while addressing them during conversation.

3. The corresponding results suggest that the mechanism outputs a directional signal to sounds ahead at calling song frequency and to sounds behind at echolocation frequencies, and that this combination of responses simplifies later neural processing in the cricket.

4. The worm has a small nervous system (302 neurons), whose neurons and connectivity pattern have been completely characterized, the neural circuit controlling chemotaxis is well known and, when implemented on a robot, it proves to be able to cope with environmental variability and noise in sensory inputs.

5. Human grasping has inspired the humanoid hand developed at Curtin University of Technology. The corresponding system presents 10 individually controllable degrees of freedom ranging from the elbow to the fingers, and is actuated through 20 McKibben air muscles each supplied by a pneumatic pressure-balancing valve that allows for proportional control to be achieved with simple and inexpensive components.

Ⅱ. Answer the following questions based on the main text in this chapter.

1. What are the five basic human senses? Can we now well replicate them all in one integrated sensor?

2. Can you talk about the natural dynamics of human walking, and its advantages when applied to biped walking robots?

3. How do Geckos freely run up walls and scurry cling ceilings?

4. What is the difference of grasping motion between human hand and octopus arm?

5. What are bioinspired learning mechanisms in robotic field?

Ⅲ. Writing.

Imagine the future of robots: will robots transform the world?

Chapter 7

Artificial Intelligence

Artificial intelligence (AI), sometimes called machine intelligence, is intelligence demonstrated by machines, in contrast to the natural intelligence displayed by humans and other animals. Some of the activities that it is designed to do is speech recognition, learning, planning and problem solving. Since robotics is the field concerned with the connection of perception to action, artificial intelligence must have a central role in robotics if the connection is to be intelligent. Artificial intelligence addresses the crucial questions of: what knowledge is required in any aspect of thinking; how should that knowledge be represented; and how should that knowledge be used. Robotics challenges artificial intelligence by forcing it to deal with real objects in the real world. In this chapter, we discuss the different definitions of artificial intelligence. We then discuss how machines learn and how a robot works in general. Finally we discuss the limitations of AI and the influence the media has on our preconceptions of AI.

7.1 Introduction to AI

Chris: Siri, should I lie about my weight on my dating profile?

Siri: I can't answer that, Chris.

Siri is not the only virtual assistant that will struggle to answer this question (see Fig.7.1).

Almost two thirds of people provide inaccurate information about their weight on dating profiles. Ignoring, for a moment, what motivates people to lie about their dating profiles, why is it so difficult, if not impossible, for digital assistants to answer this question?

To better understand this challenge, it is necessary to look behind the scene and to see how this question is processed by Siri. First, the phone's microphone needs to translate the changes in air pressure (sounds) into a digital signal that can then be stored as data in the memory of the phone. Next, this data needs to be sent through the internet to a powerful computer in the cloud. This computer then tries to classify the sounds recorded into written words. Afterwards, an artificial intelligence system needs to extract the meaning of this combination of words. Notice that it even needs to be able to pick the right meaning for the homophone "lie". Chris does not want to lie down on his dating profile, he is wondering if he should put inaccurate information on it.

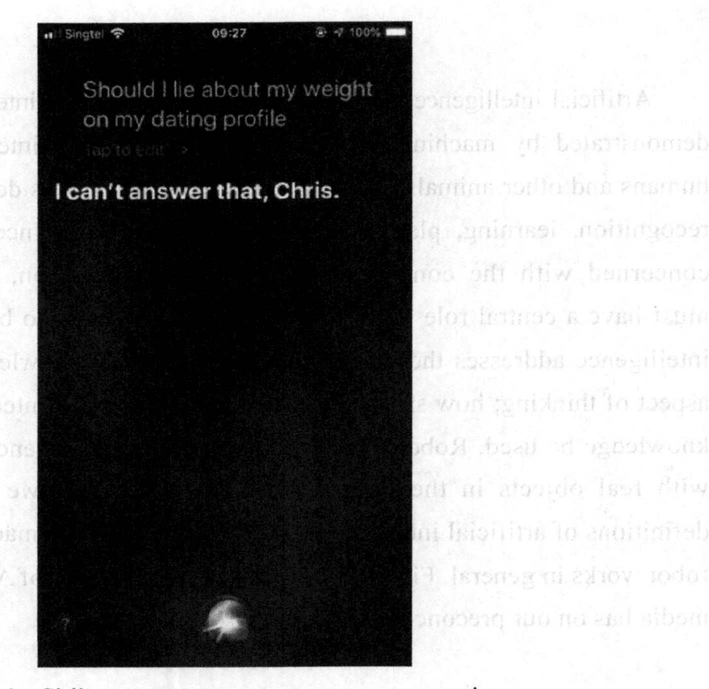

Fig.7.1 Siri's response to a not so uncommon question

While the above steps are difficult and utilize several existing AI techniques, the next step is one of the hardest. Assuming Siri fully understands the meaning of Chris's question, what advice should Siri give? To give the correct advice, it would need to know what a person's weight means and how the term relates to their attractiveness. Siri needs to know that the success of dating depends heavily on both participants considering each other attractive—and that most people are motivated to date. Furthermore, Siri needs to know that online dating participants cannot verify the accuracy of information provided until they meet in person. Siri also needs to know that honesty is another attribute that influences attractiveness. While deceiving potential partners online might make Chris more attractive in the short run, it would have a negative effect once Chris meets his date face-to-face.

But this is not all. Siri also needs to know that most people provide inaccurate information on their online profiles, and that a certain amount of dishonesty is not likely to impact Chris's long-term attractiveness with a partner. Siri should also be aware that women select only a small portion of online candidates for first dates and that making this first cut is essential for having any chance at all of convincing the potential partners of Chris's other endearing qualities.

There are many moral approaches that Siri could be designed to take. Siri could take a consequentialist approach. This is the idea that the value of an action depends on the consequences it has. The best known version of consequentialism is the classical utilitarianism of Jeremy Bentham and John Stuart Mill. These philosophers would no doubt advise Siri to maximize happiness: not just Chris's happiness, but also the happiness of his prospective date. So, on the consequentialist approach, Siri might give Chris advice that would maximize his chances to not only have many first dates, but maximize the chances for Chris to find true love.

Alternatively, Siri might be designed to take a deontological approach. A deontologist like Immanuel Kant might prioritise duty over happiness. Kant might advise Chris that lying is wrong. He has a duty not to lie so he should tell the truth about his weight, even if this would decrease his chances of getting a date.

A third approach Siri could take would be a virtue ethics approach. Virtue ethics tend to see morality in terms of character. Aristotle might advise Chris that his conduct has to exhibit virtues such as honesty.

Lastly, Siri needs to consider whether it should give a recommendation at all. Providing wrong advice might damage Siri's relationship to Chris and he might consider switching to another phone with another digital assistant.

This little example shows that questions that seem trivial on the surface might be very difficult for a machine to answer. Not only do these machines need the ability to process sensory data, they also need to be able to extract the correct meaning from it and then represent this meaning in a data structure that can be digitally stored. Next, the machine needs to be able to process the meaning and conclude with desirable actions. This whole process requires knowledge about the world, logical reasoning and skills to learn and adapt. Having these abilities may make the machine autonomous.

There are various definitions of "autonomy" and "autonomous" in AI, robotics and ethics. At its simplest, autonomous simply refers to the ability of a machine to operate for a period of time without a human operator. Exactly what that means differs from application to application. What is considered "autonomous" in a vehicle is different to what is considered "autonomous" in a weapon. In bioethics, autonomy refers to the ability of humans to make up their own minds about what treatment to accept or refuse.

On the first of these definitions, Siri is an autonomous agent that attempts to answer spoken questions. Some questions Siri tries to answer require more intelligence, meaning more background, reasoning ability and knowledge, than others. The paragraphs that follow define and describe the characteristics that make something artificially intelligent and an agent.

The field of artificial intelligence has evolved from humble beginnings to a field with global impact. The definition of AI and of what should and should not be included has changed over time. Experts in the field joke that AI is everything that computers cannot currently do. Although facetious on the surface, there is a sense that developing intelligent computers and robots means creating something that does not exist today. Artificial intelligence is a moving target.

Indeed, even the definition of AI itself is volatile and has changed over time. Some people define AI as "a system's ability to correctly interpret external data, to learn from such data, and to use those learnings to achieve specific goals and tasks through flexible adaptation". Some researchers define AI as "the field that studies the synthesis and analysis of computational agents that act intelligently". An agent is something (or someone) that acts. An agent is intelligent when:

① Its actions are appropriate for its circumstances and its goals;

② It is flexible to changing environments and changing goals;

③ It learns from experience;

④ It makes appropriate choices given its perceptual and computational limitations.

Some people define AI as "the study of (intelligent) agents that receive precepts from the environment and take action. Each such agent is implemented by a function that maps percepts to actions, and we cover different ways to represent these functions, such as production systems, reactive agents, logical planners, neural networks, and decision-theoretic systems".

They also identify four schools of thought for AI and focus on creating machines that think like humans. Research within this school of thought seeks to reproduce, in some manner, the processes, representations, and results of human thinking on a machine. A second school focuses on creating machines that act like humans. It focuses on action, what the agent or robot actually does in the world, not its process for arriving at that action. A third school focuses on developing machines that act rationally. Rationality is closely related to optimality. These artificially intelligent systems are meant to always do the right thing or act in the correct manner. Finally, the fourth school is focused on developing machines that think rationally. The planning and/or decision-making that these machines will do is meant to be optimal. Optimal here is naturally relevant to some problems that the system is trying to solve.

We have provided three definitions. Perhaps the most basic element common to all of them is that AI involves the study, design and building of intelligent agents that can achieve goals. The choices an AI makes should be appropriate to its perceptual and cognitive limitations. If an AI is flexible and can learn from experience as well as sense, plan and act on the basis of its initial configuration, it might be said to be more intelligent than an AI that just has a set of rules that guides a fixed set of actions. However, there are some contexts in which you might not want the AI to learn new rules and behaviors, during the performance of a medical procedure, for example. Proponents of the various approaches tend to stress some of these elements more than others. For example, developers of expert systems see AI as a repository of expert knowledge that humans can consult, whereas developers of machine learning systems see AI as something that might discover new knowledge. As we shall see, each approach has strengths and weaknesses.

AI currently works best in constrained environments, but has trouble with open worlds, poorly defined problems, and abstractions. Constrained environments include simulated environments and environments in which prior data accurately reflects future challenges. The real world, however, is open in the sense that new challenges arise constantly. Humans use solutions to prior related problems to solve new problems. AI systems have limited ability to reason analogically from one situation to another, and thus tend to have to learn new solutions even for closely related problems. In general, they lack the ability to reason abstractly about problems and to use common sense to generate solutions to poorly defined problems.

The ultimate hypothetical goal is achieving superintelligence (ASI) which is far surpassing that of the brightest and most gifted human minds. Due to recursive self-improvement, superintelligence is expected to be a rapid outcome of creating artificial general intelligence.

7.2 Achieving AI

There are many ways of achieving AI, some of them are listed in Fig.7.2.

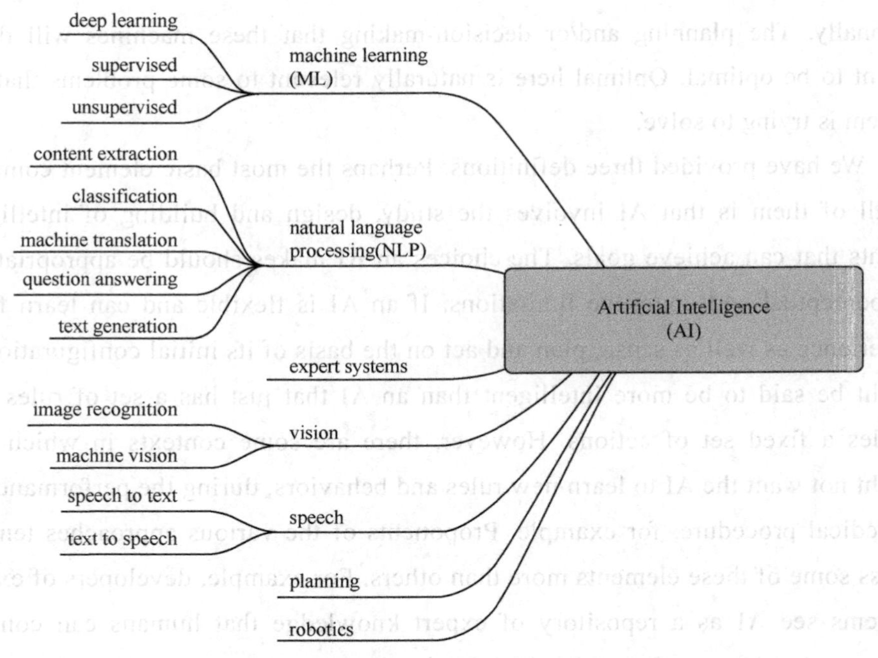

Fig.7.2 AI branches

7.3 Machine Learning

Machine learning (ML) is a sub-field of AI focused on the creation of algorithms that use experience with respect to a class of tasks and feedback in the form of a performance measure to improve their performance on that task. Contemporary machine learning is a sprawling, rapidly changing field.

According to Arthur Samuel, machine learning is defined as the field of study that gives computers the ability to learn without being explicitly programmed. Arthur Samuel was famous for his checkers playing program. Machine learning is used to teach machines how to handle the data more efficiently. Sometimes after viewing the data, we cannot interpret the extract information from the data. In that case, we apply machine learning. With the abundance of datasets available, the demand for machine learning is in rise. Many industries apply machine learning to extract relevant data. The purpose of machine learning is to learn from the data. Many studies have been done on how to make machines learn by themselves without being explicitly programmed. Many mathematicians and programmers apply several approaches to find the solution of this problem which are having huge data sets.

Machine learning relies on different algorithms to solve data problems. Data scientists like to point out that there's no single one-size-fits-all type of algorithm that is best to solve a problem. The kind of algorithm employed depends on the kind of problem you wish to solve, the number of variables, the kind of model that would suit it best and so on. Here's a brief introduction at some of the commonly used algorithms in machine learning.

7.3.1 Supervised Learning

It centers on methods such as regression and classification. To solve a classification problem, experiences in the form of data are labelled with respect to some target categorization. The labelling process is typically accomplished by enlisting the effort of humans to examine each piece of data and to label the data. For supervised learning classification problems, performance is measured by calculating the true positive rate (the ratio of the true positives over all positives, correctly labelled or not) and the false positive rate (the ratio of false positives over all negatively classified data, correctly and incorrectly labelled). The result of this machine learning process is called a classifier. A classifier is software that can automatically predict the label of a new piece of data. A machine learning classifier that categorizes labelled data with a true positive rate of 100% and a false positive rate of 0% is a perfect classifier. The supervised learning process then is the process by which unlabelled data is fed to a developing classifier, and over the course of working through some training data, the classifier's performance improves. Testing the classifier requires the use of a second label data-set called the test data set. In practice, often one overall data-set is carved into a training and test set on which the classifier is then trained and tested. The testing and training process may be time-consuming, but once a classifier is created, it can be used to quickly categorize incoming data. (Fig.7.3)

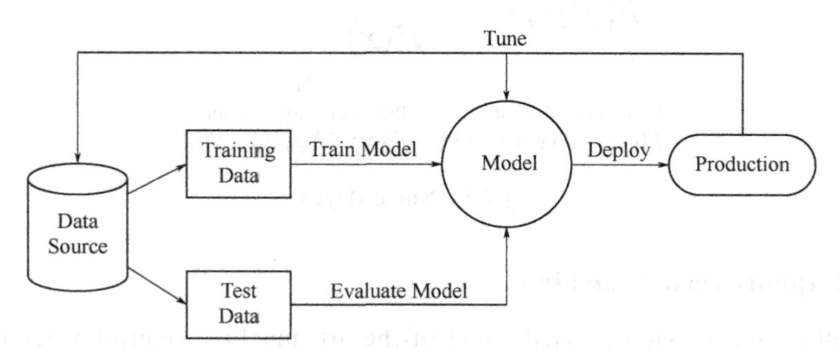

Fig.7.3 Supervised learning workflow

(1) Decision tree

Decision tree (Fig.7.4) is a graph to represent choices and their results in form of a tree. The nodes in the graph represent an event or choice, and the edges of the graph represent the decision rules or conditions. Each tree consists of nodes and branches. Each node represents attributes in a group that is to be classified, and each branch represents a value that the node can take.

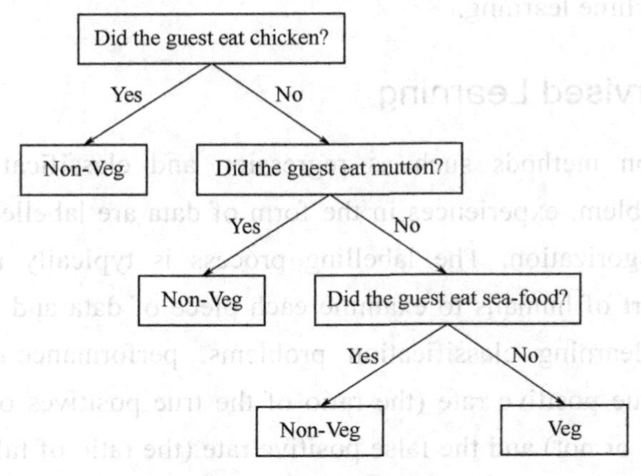

Fig.7.4 Decision tree

(2) Naive Bayes

It is a classification technique based on Bayes theorem with an assumption of independence among predictors. In simple terms, a Naive Bayes classifier assumes that the presence of a particular feature in a class is unrelated to the presence of any other feature. Naive Bayes mainly targets the text classification industry. It is mainly used for clustering and classification purposes depends on the conditional probability of happening. (Fig.7.5)

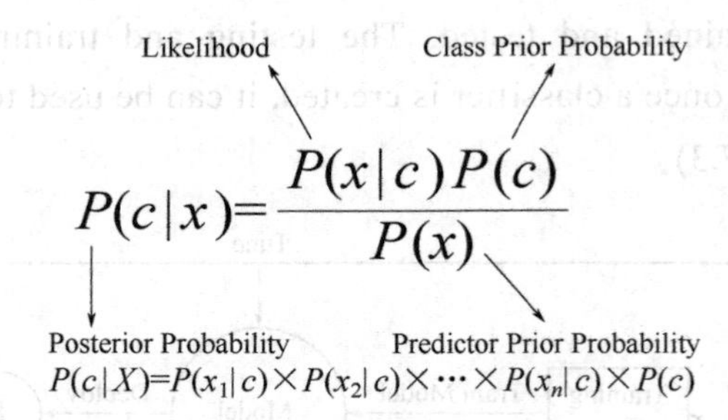

$$P(c|x)=\frac{P(x|c)P(c)}{P(x)}$$

$$P(c|X)=P(x_1|c)\times P(x_2|c)\times\cdots\times P(x_n|c)\times P(c)$$

Fig.7.5 Naive Bayes

(3) Support vector machine

Another most widely used state-of-the-art machine learning technique is support vector machine (SVM) (Fig.7.6). In machine learning, support vector machines are supervised learning models with associated learning algorithms that analyze data used for classification and regression analysis. In addition to performing linear classification, SVMs can efficiently perform a non-linear classification using what is called the kernel trick, implicitly mapping their inputs into high-dimensional feature spaces. It basically draw margins between the classes. The margins are drawn in such a fashion that the distance between the margin and the classes is maximum and hence, minimizing the classification error.

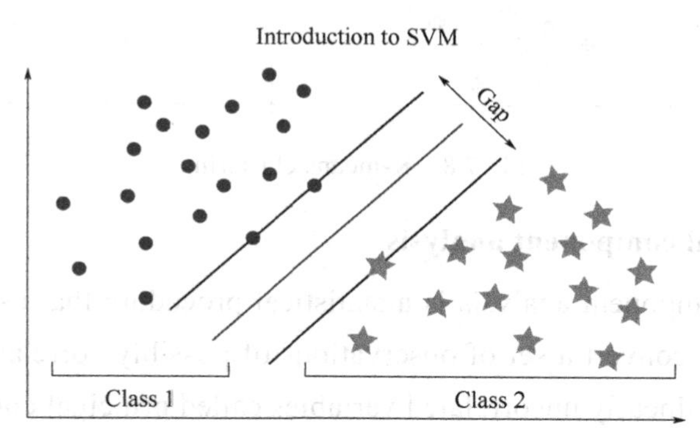

Fig.7.6 Support vector machine

7.3.2 Unsupervised Learning

It is more focused on understanding data patterns and relations than on prediction (Fig.7.7). It involves methods such as principal components analysis and clustering. These are often used as exploratory precursors to supervised learning methods.

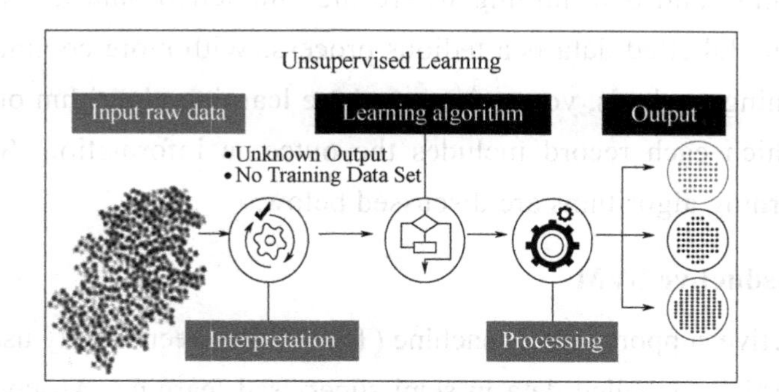

Fig.7.7 Unsupervised learning

(1) K-means clustering (Fig.7.8)

K-means is one of the simplest unsupervised learning algorithms that solve the well-known clustering problem. The procedure follows a simple and easy way to classify a given data set through a certain number of clusters. The main idea is to define k centers, one for each cluster. These centers should be placed in a cunning way because different location causes different result. So, the better choice is to place them as much as possible far away from each other.

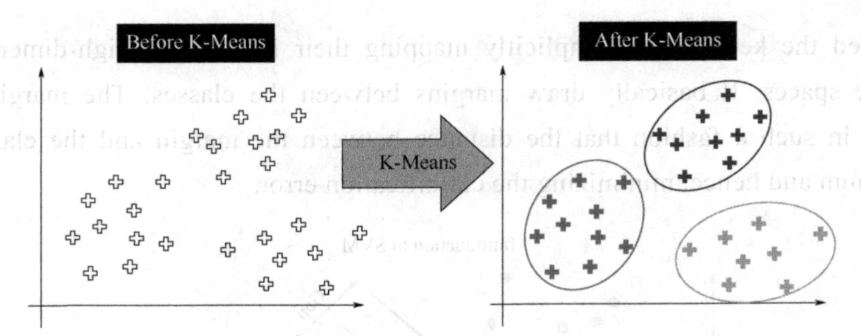

Fig.7.8 K-means clustering

(2) Principal component analysis

Principal component analysis is a statistical procedure that uses an orthogonal transformation to convert a set of observations of possibly correlated variables into a set of values of linearly uncorrelated variables called principal components. In this, the dimension of the data is reduced to make the computations faster and easier. It is used to explain the variance-covariance structure of a set of variables through linear combinations. It is often used as a dimensionality-reduction technique.

7.3.3 Semi Supervised Learning

Semi supervised machine learning is a combination of supervised and unsupervised machine learning methods. It can be fruit-full in those areas of machine learning and data mining where the unlabelled data is already present and getting the labelled data is a tedious process. With more common supervised machine learning methods, you train a machine learning algorithm on a "labelled" dataset in which each record includes the outcome information. Some of semi supervise learning algorithms are discussed below.

(1) Transductive SVM

Transductive support vector machine (TSVM) has been widely used as a means of treating partially labelled data in semi supervised learning. Around it, there has been mystery because of lack of understanding of its foundation in generalization. It is used to label the unlabelled data in such a way that the margin is maximum between the labelled and unlabelled data. Finding an exact solution by TSVM is a NP-hard problem.

(2) Generative models

A generative model is the one that can generate data. It models both the features and the class (i.e. the complete data). If we model $P(x,y)$, we can use this probability distribution to generate data points, and hence all algorithms modeling

$P(x,y)$ are generative. One labelled example per component is enough to confirm the mixture distribution.

(3) Self-training

In self-training, a classifier is trained with a portion of labelled data. The classifier is then fed with unlabelled data. The unlabelled points and the predicted labels are added together in the training set. This procedure is then repeated further. Since the classifier is learning itself, hence the name is self-training.

7.3.4 Reinforcement Learning

It is a third type of machine learning. Reinforcement learning (Fig.7.9) does not focus on the labelling of data, but rather attempts to use feedback in the form of a reinforcement function to label states of the world as more or less desirable with respect to some goal. Consider, for example, a robot attempting to move from one location to another. If the robot's sensors provide feedback telling it its distance from a goal location, then the reinforcement function is simply a reflection of the sensor's readings. As the robot moves through the world, it arrives at different locations which can be described as states of the world. Some world states are more rewarding than others. Being close to the goal location is more desirable than being further away or behind an obstacle. Reinforcement learning learns a policy, which is a mapping from the robot's action to expected rewards. Hence, the policy tells the system how to act in order to achieve the reward.

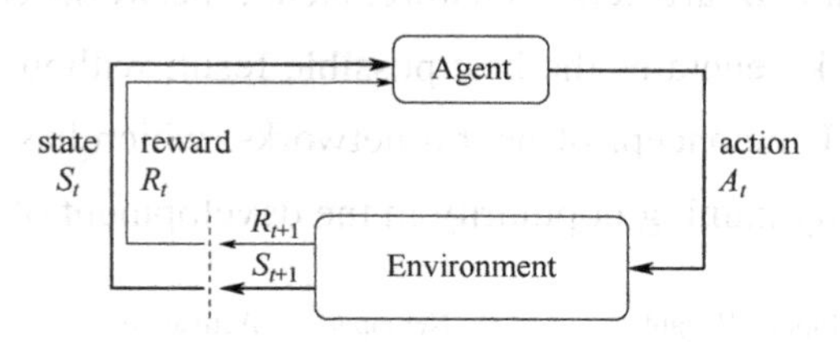

Fig.7.9 Reinforcement learning

(1) Multitask learning

Multitask learning is a sub-field of machine learning that aims to solve multiple different tasks at the same time, by taking advantage of the similarities between different tasks. This can improve the learning efficiency and also act as a regularization. Formally, if there are n tasks (conventional deep learning approaches aim to solve just 1 task using 1 particular model), where these n tasks or a subset of them are related to each other but not exactly identical, multitask learning (MTL) will help in improving the learning of a particular model by using the knowledge contained in all the n tasks.

(2) Ensemble learning

Ensemble learning is the process by which multiple models, such as classifiers or experts, are strategically generated and combined to solve a particular computational intelligence problem. Ensemble learning is primarily used to improve the performance of a model, or reduce the likelihood of an unfortunate selection of a poor one. Other applications of ensemble learning include assigning a confidence to the decision made by the model, selecting optimal features, data fusion, incremental learning, non-stationary learning and error-correcting.

(3) Boosting

The term "Boosting" refers to a family of algorithms which converts weak learner to strong learners. Boosting is a technique in ensemble learning which is used to decrease bias and variance. Boosting is based on the question posed by Kearns and Valiant: "Can a set of weak learners create a single strong learner?" A weak learner is defined to be a classifier, and a strong learner is a classifier that is arbitrarily well-correlated with the true classification.

(4) Neural networks

A neural network is a series of algorithms that endeavors to recognize underlying relationships in a set of data through a process that mimics the way the human brain operates (Fig.7.10). In this sense, neural networks refer to systems of neurons, either organic or artificial in nature. Neural networks can adapt to changing input; so the network generates the best possible result without needing to redesign the output criteria. The concept of neural networks, which has its roots in artificial intelligence, is swiftly gaining popularity in the development of trading systems.

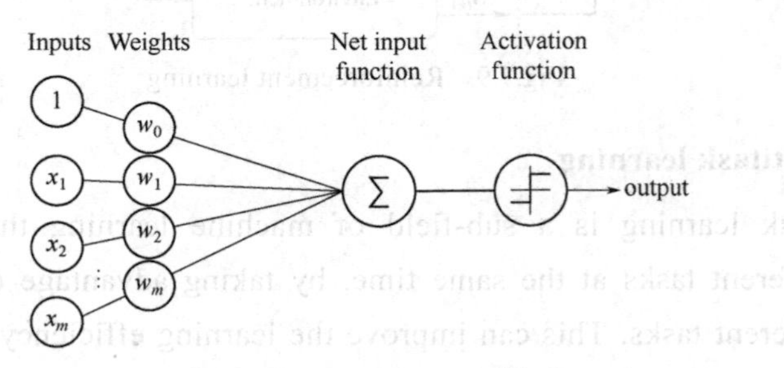

Fig.7.10 Neural networks

An artificial neural network behaves the same way. It works on three layers. The input layer takes input. The hidden layer processes the input. Finally, the output layer sends the calculated output.

(5) Supervised neural network

In the supervised neural network, the output of the input is already known. The predicted output of the neural network is compared with the actual output. Based on the error, the parameters are changed, and then fed into the neural network again. Supervised neural network is used in feed forward neural network. (Fig.7.11)

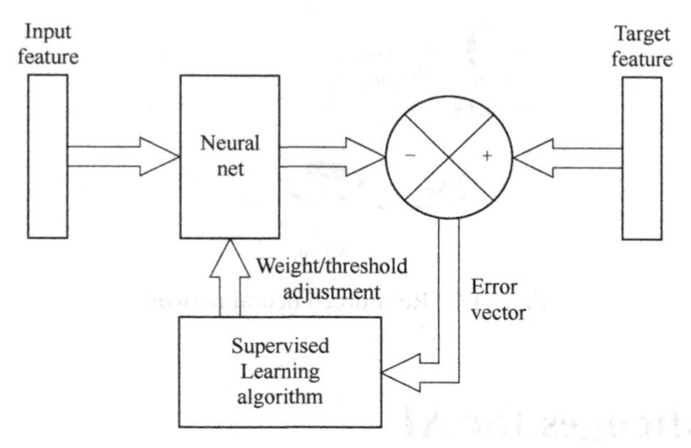

Fig.7.11 Supervised neural network

(6) Unsupervised neural network

The neural network has no prior clue about the output or the input. The main job of the network is to categorize the data according to some similarities. The neural network checks the correlation between various inputs and groups them. (Fig.7.12)

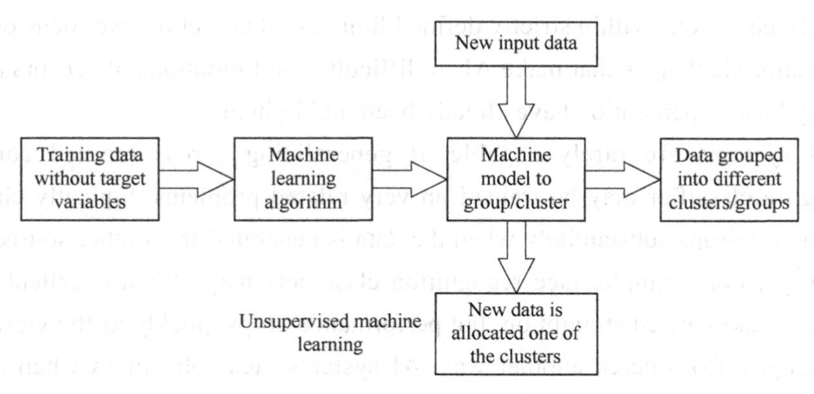

Fig.7.12 Unsupervised neural network

(7) Reinforced neural network

Reinforcement neural learning refers to goal-oriented algorithms, which learn how to attain a complex objective (goal) or maximize along a particular dimension over many steps; for example, maximize the points won in a game over many moves. They can start from a blank slate, and under the right conditions they achieve superhuman performance. Like a child incentivized by spankings and candy, these algorithms are penalized when they make the wrong decisions and rewarded when they make the right ones—this is reinforcement. (Fig.7.13)

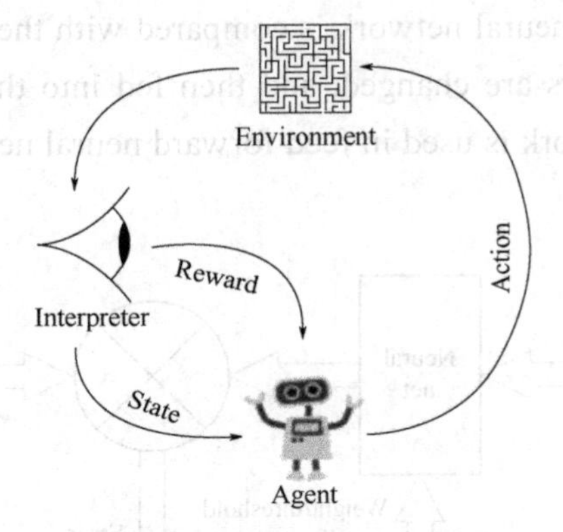

Fig.7.13 Reinforced neural network

7.4 Challenges for AI

The sections above have hinted at why AI is hard. It should also be mentioned that not all software is AI. For example, simple sorting and search algorithms are not considered intelligent. Moreover, a lot of non-AI is smart. For example, control algorithms and optimization software can handle everything from airline reservation systems to the management of nuclear power plants. But they only take well-defined actions within strictly defined limits. In this section, we focus on some of the major challenges that make AI so difficult. The limitations of sensors and the resulting lack of perception have already been highlighted.

AI systems are rarely capable of generalizing across learned concepts. Although a classifier may be trained on very related problems, typically classifier performance drops substantially when the data is generated from other sources or in other ways. For example, face recognition classifiers may obtain excellent results when faces are viewed straight on, but performance drops quickly as the view of the face changes. Considered another way, AI systems lack robustness when dealing with a changing, dynamic, and unpredictable world. As mentioned, AI systems lack common sense. Put another way, AI systems lack the enormous amount of experience and interactions with the world that constitute the knowledge that is typically called common sense. Not having this large body of experience makes even the most mundane task difficult for a robot to achieve. Moreover, lack of experience in the world makes communicating with a human and understanding a human's directions difficult. This idea is typically described as common ground.

Although a number of software systems have claimed to have passed the

Turing test, these claims have been disputed. No AI system has yet achieved strong AI, but some may have achieved weak AI based on their performance on a narrow, well defined task (like beating a grandmaster in chess or Go, or experienced players in Poker). Even if an AI agent is agreed to have passed the Turing test, it is not clear whether the passing of the test is a necessary and sufficient condition for intelligence.

AI has been subject to many hype cycles. Often even minor advancements have been hailed as major breakthroughs with predictions of soon to come autonomous intelligent products. These advancements should be considered with respect to the narrowness of the problem attempted. For example, early types of autonomous cars capable of driving thousands of miles at a time (under certain conditions) were already being developed in the 1980s in the US and Germany. It took, however, another 30+ years for these systems to just begin to be introduced in non-research environments. Hence, predicting the speed of progression of AI is very difficult—and in this regard, most prophets have simply failed.

7.5 Application Areas of AI

It is starting to become possible to merge humans and machines. Robotic components that can replace biological anatomy are under active development. No ethical dilemma ensues when the use of robotic replacement parts is for restorative purposes. Robotic restoration takes place when a missing physical or cognitive capability is replaced with an equally functional mechanical or electronic capability. The most common example of robotic restoration is the use of robotic prosthetics. Enhancement, on the other hand, occurs when a physical or cognitive capability is replaced with an amplified or improved mechanical or electronic capability.

It is not always clear when a restorative prosthetic becomes an enhancement. Technology is changing rapidly and prosthetics are becoming so advanced as to allow for significant increases in some abilities. Moreover, while some functions may only be restored, others may be enhanced. Typically, these distinctions do not inherently result in ethical dilemmas. Non-invasive mechanical and electrical devices, such as exo-skeletons, are also being developed to make people stronger, faster, or more capable in some ways. These systems may potentially be purchased by people for the sake of enhancement.

Healthcare is another application of AI and robotics that raises ethical issues. Robot have been proposed for a wide variety of roles in healthcare including assisting older adults in assisted living, assisting with rehabilitation, surgery, and delivery. Currently robot-assisted surgery is the predominant application of robots within the healthcare industry. Robots are also being developed to deliver items in a hospital environment and for using ultraviolet light to disinfect hospital and surgical rooms.

7.5.1 Robots and Telemedicine

Robots have been suggested as an important method for performing telemedicine whereby doctors perform examinations and determine treatments of patients from a distance (see Fig.7.14). The use of robots for telemedicine offers both benefits and risks. This technology may afford a means for treating distantly located individuals that would otherwise only be able to see a doctor under extreme circumstances. Telemedicine may thus encourage patients to see the doctor more often. It may also decrease the cost of providing healthcare to rural populations. On the negative side, the use of telemedicine may result in and even encourage a substandard level of healthcare, when being used in an exaggerated way. It might also result in the misdiagnosis of certain ailments which are not easily evaluated remotely.

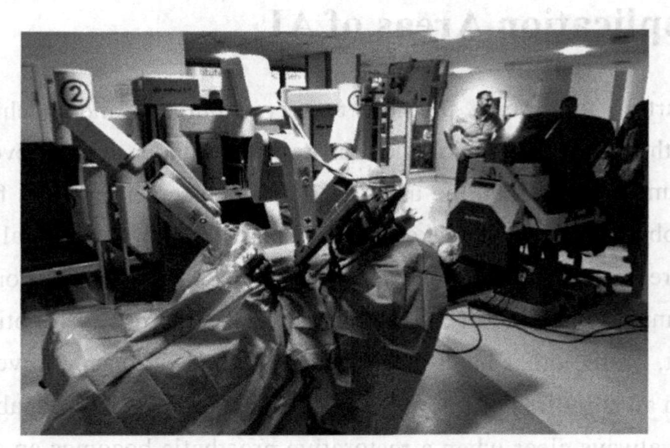

Fig.7.14 Da Vinci surgical system (Source: Cmglee)

Robots are also being introduced as a benefit to older adults to combat social isolation. Social isolation occurs for a variety of reasons such as children entering adulthood and leaving the home, friends and family ageing and passing away. Older adults that reside in nursing homes may feel increasingly isolated which can result in depression.

Researchers have developed robots, such as Paro, in an attempt to reduce feelings of loneliness and social isolation in these older adults. Ethical concerns about the use of the Paro robot (see Fig.7.15) have been raised. The main concerns are that patients with dementia may not realize that the robot is a robot, even if they are told (whatever the consequences may be). Moreover, the use of the robot may further increase actual social isolation by reducing the incentive of family members to visit. Yet, for this use case, many would argue that the benefits clearly outweigh the concerns.

Fig.7.15 Paro robot (Source: National Institute of Advanced Industrial Science and Technology)

Perhaps more controversial thing is the use of AI and robotics to provide encouraging nudges to push patients towards a particular behavioral outcome. Robotic weight loss coaches, for example, have been proposed and developed that ask people about their eating habits and remind them to exercise. These robots are meant to help people stick to diets, but ethical concerns arise related to the issue of autonomy. Specifically, people should have the autonomy to choose how they want to live and not be subjected to the influence of an artificially intelligent system. These systems also raise issues related to psychological manipulation if their interactions are structured in a way that is known to be most influential. A variety of methods exist, such as the foot-in-the-door technique which could be used to manipulate a person.

Recently, artificial systems have been proposed as a means for performing preliminary psychological evaluations. Ideally, these systems could be used to detect depression from online behavior and gauge whether or not a treatment intervention is required. The use of such systems offers a clear benefit in that, by identifying individuals at risk, they may be able to prevent suicides or other negative outcomes. On the other hand, these systems still raise questions of autonomy and the potential development of nanny technologies which prevent humans from working through their own problems unfettered. Along a similar line of reasoning, virtual agents have been developed for interviewing Post Traumatic Stress Disorder (PTSD) suffers. Research has shown that individuals with PTSD are more likely to open up to a virtual agent than to a human therapist.

Looking at one final technology, exo-skeletons have been developed to assist individuals with lower-limb disabilities. These systems are most often used for rehabilitation and training. Recently they have allowed paraplegic patients to stand and take small steps. Ethical issues arise when these systems are incorrectly viewed as a cure for disease rather than a tool.

Overall, robotics and AI have the potential to revolutionize healthcare. These technologies may well provide lasting benefits in many facets of care ranging from surgery to diagnose disease. Certainly, a society must carefully analyze the benefits and costs of these systems. For instance, computer aided detection of cancer affects decisions in complex ways. Some researchers examined the quality of decisions that result when healthcare providers use computer-aids to detect cancer in mammograms. They found that the technology helped more novice mammogram readers but hindered more experienced readers. They note that this differential effect, even if subtle, may be clinically significant. The authors suggest that detection algorithms and protocols should be developed that include the experience of the user in the type of decision support. In 2019, a study on more than 9,400 women published by the Journal of the National Cancer Institute found that AI is overwhelmingly better in detecting pre-cancerous cells than human doctors.

7.5.2 Education

In education, AI systems and social robots have been used in a variety of contexts. Online courses are widely used. For example, the University of Phoenix is now (technically) one of the largest universities in the world—since hundreds of thousands of students are enrolled in their online courses. In such a highly digital learning environment, it is much easier to integrate AI that helps students not only with their administrative tasks, but also with their actual learning experiences.

AI systems, such as Amelia from IPSoft, may one day advise students on their course selecting and provide general administrative support. This is not fundamentally different from other chatbot platforms such as Amazon's Lex, Microsoft's Conversation or Google's Chatbase. They all provide companies and organizations with tools to create their own chatbot that users can interact with on their respective websites or even on dedicated messaging platforms. While these bots may be able to provide some basic support, they do have to fall back to a human support agent when encountering questions that go beyond the knowledge stored in their databases.

Another form of supporting education with AI from an organizational perspective is plagiarism checking. In an age where students are able to copy and paste essays easily from material found online, it is increasingly important to check if the work submitted is truly the student's original work or just a slightly edited Wikipedia article. Students are of course aware of their teachers' ability to google the texts in their essays, and therefore are aware that they need to do better than a plain copy and paste. Good plagiarism checking software goes far beyond matching identical phrases and is able to detect similarities and approximations, even search for patterns in the white space. Artificial intelligence is able to detect the similarities of text patterns and empowers teachers to quickly check student work against all major sources on the internet including previously submitted student contributions that were never openly published.

AI systems have several advantages over human teachers that make them attractive for online learning. First, they are extremely scalable. Each student can work with a dedicated AI system which can adapt the teaching speed and difficulty to the students' individual needs. Second, such a system is available at any time, for an unconstrained duration at any location. Moreover, such an agent does not get tired and does have an endless supply of patience. Another advantage of AI teaching systems is that students might feel less embarrassed. Speaking a foreign language to a robot might be more comfortable for a novice speaker.

The autonomous teaching agents work best in constrained topics, such as math, in which good answers can be easily identified. Agents will fail in judging the beauty of a poem or appreciate the novelty in thought or expression.

Another teaching context in which robots and AI systems show promising results is teaching children with special needs, more specifically children with autism spectrum disorder. The robots' limited expressivity combined with its repetitive behavior (that is perceived by many as boring) is in this context actually a key advantage. The use of robots for general purpose childcare is ethically questionable.

Key vocabularies

agent [ˈeɪdʒənt] ***n.*** 代理人，代理商；药剂；行为主体；动因；施动者
ailment [ˈeɪlmənt] ***n.*** 疾病，病痛；不安
algorithm ***n.*** 算法；计算程序
autism spectrum disorder ***n.*** 自闭症谱系障碍
bot [bɒt] ***n.*** 机器人程序；机器人；马蝇的幼虫
common ground ***n.*** 共同基础；一致之处
consequentialism ***n.*** 效果论

consequentialist *n.* 后果论者，结果论者
dementia [dɪˈmenʃə] *n.* 痴呆，精神错乱
deontological [diːɔntəˈlɔdʒikəl] *adj.* 义务论式的
deontologist *n.* 道义论学家；义务论学家
dispute [dɪˈspjuːt] *n.* 争论，辩论；纠纷. *v.* 对……质疑，否认；争论，辩论
enlist [ɪnˈlɪst] *v.* 争取，谋取；赞助，支持
facet [ˈfæsɪt] *n.* 部分，方面；（构成昆虫或甲壳动物复眼的）小眼面
foot-in-the-door *n.* 成功的第一步，成功的开始
fusion [ˈfjuːʒn] *n.* 融合；熔接；结合
gauge [geɪdʒ] *n.* 测量标准；测量仪器；评估；胎压计. *vt.* 测量；评估，判断；采用
homophone [ˈhɒməfəʊn] *n.* 同音异义词
incentive [ɪnˈsentɪv] *n.* 动机；刺激；诱因；鼓励
incremental [ˌɪŋkrəˈmentl] *adj.* 增加的；递增的
mammogram [ˈmæməgræm] *n.* 乳房X线照片
mimics [ˈmɪmɪks] *v.* 模仿(人的言行举止)；(外表或行为举止)像. *adj.* 模仿的；拟态的
mundane [mʌnˈdeɪn] *adj.* 单调的，平凡的，平淡的. *n.* 平凡的事物
novice [ˈnɒvɪs] *n.* 初学者，新手
nudge [nʌdʒ] *n.* 轻推，碰；说服. *v.* 劝说，鼓励
paraplegic [ˌpærəˈpliːdʒɪk] *n.* 截瘫病人，下身麻痹患者. *adj.* 截瘫的，下身瘫痪的
plagiarism [ˈpleɪdʒərɪzəm] *n.* 剽窃；抄袭；剽窃物；抄袭物
precept [ˈpriːsept] *n.* 规则；格言；训诫；命令
proponent [prəˈpəʊnənt] *n.* 支持者，拥护者；提倡者
recursive [rɪˈkɜːsɪv] *adj.* 回归的，递归的；循环的
rehabilitation [ˌriːəˌbɪlɪˈteɪʃn] *n.* 康复，复原；恢复；修复，翻新
repository [rɪˈpɒzətri] *n.* 仓库；贮藏室；自然资源丰富的地区；数据库
slate [sleɪt] *n.* 板岩，石板；石板瓦. *adj.* 石板色的，深蓝灰色的
telemedicine [ˈtelɪˌmedɪsɪn] *n.* 远程医疗，远距离治病
therapist [ˈθerəpɪst] *n.* 治疗专家；特定疗法技师
transductive *adj.* 直推式的
unfetter [ʌnˈfetəd] *adj.* 不受限制的，无拘无束的. *v.* 解开脚链；使自由，解放

Exercises

Ⅰ. Translate the following English into Chinese.

1. Machine learning is a sub-field of AI focused on the creation of algorithms that use experience with respect to a class of tasks and feedback in the form of a

performance measure to improve their performance on that task.

2. AI systems are rarely capable of generalizing across learned concepts. Although a classifier may be trained on very related problems, typically classifier performance drops substantially when the data is generated from other sources or in other ways.

3. Currently robot-assisted surgery is the predominant application of robots within the healthcare industry. Robots are also being developed to deliver items in a hospital environment and for using ultraviolet light to disinfect hospital and surgical rooms.

4. Artificial intelligence is able to detect the similarities of text patterns and empowers teachers to quickly check student work against all major sources on the internet including previously submitted student contributions that were never openly published.

5. The robots' limited expressivity combined with its repetitive behavior (that is perceived by many as boring) is in this context actually a key advantage.

Ⅱ. Answer the following questions based on the main text in this chapter.

1. What is deep learning and how does it work?

2. Please find more AI scenes, except those introduced in the text.

3. Can you explain why AI has advantages over human doctors in pre-cancerous detection?

4. Is there any disadvantage of AI?

5. How will AI change the world?

Ⅲ. Writing.

The beauty and danger of AI.

References

[1] Bartneck C, Lütge C, Wagner A, et al. An Introduction to ethics in robotics and AI [M]. Cham: Springer, 2021: 5-16.

[2] Fayemi P E, Maranzana N, Aoussat A, et al. Bio-inspired design characterisation and its links with problem solving tools[C]//Marjanović D, Štorga M, Pavković N, et al. Proceedings of International Design Conference-Design 2014. Glasgow, the Design Society, 2014: 173-182.

[3] Knippers J, Nickel K G, Speck T. Biomimetic research for architecture and building construction[M]. Cham: Springer, 2016.

[4] Lenau T A, Metzea A-L, Hesselberg T. Paradigms for biologically inspired design[C]//A Lakhtakia.Proceedings of SPIE (Vol. 10593). SPIE - International Society for Optical Engineering, 2018.

[5] Meyer J-A, Guillot A. Biologically-inspired robots[M]. Cham: Springer, 2008.

[6] Neurohr R, Dragomirescu C. Bionics in engineering-defining new goals in engineering education at "Politehnica" University of Bucharest[C]. International Conference on Engineering Education—ICEE, 2007, Coimbra, Portugal.

[7] Persiani S. Biomimetics of motion: nature-inspired parameters and schemes for kinetic design [M]. Cham: Springer, 2019.

[8] Shelley T. Worms show way to efficiently move soil[OL]. https://www.eurekamagazine.co.uk/content/technology/worms-show-way-to-efficiently-move-soil. Accessed on 9/2/2022.

[9] Steinbuch R, Gekeler S. Bionic optimization in structural design: stochastically based methods to improve the performance of parts and assemblies [M]. Cham: Springer, 2016.

[10] Wolff J O, Wells D, Reid C R, et al. Clarity of objectives and working principles enhances the success of biomimetic programs[J]. Bioinspiration & Biomimmetics, 2017, 12: 051001.

[11] Wegs U G K, Bai H, Saiz E, et al. Bioinspired structural materials [J]. Nature materials, 2015, 14: 23-36.